Fragile Web

what next for nature?

Edited by Jonathan Silvertown

AUTHORS

Joanna Freeland, William Gosling, Tim Halliday, Donal O'Donnell,
Callum Roberts, David Robinson, Phil Sarre,
Jonathan Silvertown and Peter Skelton

The University of Chicago Press
Chicago and London

Acknowledgments

The editor and authors are indebted to Caroline Pond for extensive comments on the manuscript. We are also grateful for comments on individual chapters to Alastair Ewing, Julie Hawkins, Rissa de la Paz, Carmen Revenga, and Chris Thomas.

Jonathan Silvertown is professor of ecology in the department of biological sciences at the Open University. He is the author of *Demons of Eden* and *An Orchard Invisible*, both published by the University of Chicago Press.

The University of Chicago Press, Chicago 60637
The University of Chicago Press, Ltd., London

Printed in China

20 19 18 17 16 15 14 13 12 11 10   1 2 3 4 5

ISBN-13: 978-0-226-75781-0 (paper)
ISBN-10: 0-226-75781-1 (paper)

First published by the Natural History Museum, Cromwell Road, London SW7 5BD
in association with The Open University, Walton Hall, Milton Keynes MK7 6AA.

CIP data is available from the Library of Congress.

∞ The paper used in this publication meets the minimum requirements of the American National Standard for Information Sciences—Permanence of Paper for Printed Library Materials, ANSI Z39.48-1992.

Contents

Introduction

HUMANS ARE A PART OF NATURE, BUT WHICH OF US can really say that we go about our daily lives conscious of this fact, or what it means? That we are part of nature's web is a truism that is taken for granted as if it were obvious and trivial; it is actually profoundly important. The 'fragile web' of this book's title is the network of relationships among species on this planet upon which we depend for our existence and our survival. Everything we eat is a product of nature - species that have been bred, farmed, fished or collected. The oxygen in the air we breathe is the product of photosynthesis by plants, and plants also help regulate our climate.

Nature in its rich variety is called biodiversity and this book explores that variety. The first section is about the Origins of biodiversity – or in other words the source of nature's richness. Life on Earth began at least 3.5 billion years ago and for most of that time life consisted only of microbes. Then the web of nature began to get more complicated and new forms arose through evolution, with animals evolving means to escape their predators and plants evolving in size in the struggle for light with their neighbours. The connections in the web that join an animal to its prey, a flower to its pollinator, or a tree to its rivals are all part of the struggle for existence that Charles Darwin recognized drives evolution and produces opportunities for new species. At intervals in life's long and chequered history there were mass extinctions that wiped out large factions of the living world. The most recent was 65 million years ago when the dinosaurs and other large reptiles met their end and mammal diversity evolved, filling the empty places left in nature's web. If there is an opportunity, evolution will seize it and our species, with our fellow mammals, was the eventual, accidental beneficiary of that long-ago reptilian Armageddon.

The reason we focus on biodiversity is twofold. First because variety, biodiversity, underpins the functioning of nature's web. The middle section of the book is about Ecology, how biodiversity functions and what it does. Nature's web is solar powered, driven by the energy that land plants and ocean phytoplankton capture from sunlight and lock away in carbon compounds in their cells. Plants and phytoplankton not only feed all life on the planet, but by capturing carbon dioxide they help control the concentration of this gas in the Earth's atmosphere. Their capacity to lock-up carbon is now overwhelmed by the amount that we are returning to the atmosphere through burning fossil fuels such as coal and oil. Hence the rise in atmospheric carbon dioxide that now threatens to heat our world to levels that will change it beyond recognition. Regulation of the climate is just one of the services that biodiversity provides to humanity.

The second reason we focus on biodiversity is, of course, because it is under threat from human actions. Hence the sub-title of the book *What next for nature?* and the subject of the last section of the book, Fate – threats and solutions. The picture is bleak.

So many species are now threatened with extinction that the fossil record of our epoch, viewed by a geologist of the distant future, could well mirror that left behind by the mass extinction that ended the reign of the reptiles. A quarter of mammals are threatened, 40% of amphibians and perhaps a quarter of land plants. All major ocean fisheries are over-exploited and many have collapsed. Life in freshwaters is threatened by pollution, over-extraction of water for human use and by invasions of non-native species that destroy the fabric of nature's web. Climate change is tugging at the links in the web, threatening to dislocate relationships with unpredictable consequences. What can we do? The evidence is clear. We need to check and then reverse rising carbon emissions. We must halt deforestation, restore habitats and fisheries, create and protect reserves for nature, find sustainable livelihoods for the poor and feed our growing population without endangering biodiversity. This is the biggest challenge humanity has ever faced, but a species called *Homo sapiens*, or 'wise human', should be up to the task.

Although the authors of this book are passionate about biodiversity and its fate, this is a book of science, not of polemic. That means that every statement made in these pages has been distilled from independent studies that have been subject to peer review - the quality control process that preserves the integrity of science. This is a book based on the latest science, but you need no science background to read it because we believe that its subject matter is too important to be left entombed in the arcane language of science. This book is not written for scientists, it is written for the people who will decide the ultimate answer to the question we have posed: *What next for nature?* Everyone.

LEFT Humans depend upon nature in so many ways, not least for the beauty it brings into our lives, but coral reefs like this one are severely threatened.

1 CHAPTER *Biodiversity and us*

YOU COME FACE-TO-FACE WITH BIODIVERSITY every day but may not realize it. Sit down to breakfast in the morning to a bowl of cereal with milk and a cup of coffee and the likelihood is that you are enjoying the products of at least a dozen species of plants and animals: the cereal alone may contain any or all of: wheat, maize, oats, barley malt, sugar cane, bee honey, nuts and oil, flavourings and colours all extracted from plants. The cows that fill your cereal bowl with milk are the domesticated descendants of wild species. They in turn fed on plants that were also domesticated and bred from wild species of grass to produce high yields. If you prefer soymilk, of course that too is a link to the biodiversity of our planet. Yoghurt may be on your breakfast menu; it is made from milk fermented with bacteria that convert the milk sugar lactose into lactic acid. The acidity inhibits the growth of many other micro-organisms and therefore helps to preserve the product as well as giving it a characteristic taste. Choose a vanilla-flavoured yoghurt and you have added something even more exotic to your breakfast because the source of vanilla is the fermented pods of an orchid. Coffee beans, whether processed before they reached you to extract the soluble essence used in instant coffee or whether you ground the beans yourself, are ambassadors of African biodiversity at your breakfast table. Coffee is the second most valuable commodity in world trade, beaten only by oil.

Your entire breakfast is a gift of nature (albeit one you have to buy) that comes in a box labelled 'Biodiversity'. The meaning of biodiversity is life's variety. So, what is inside the box? First, there are all the species and their various breeds. All living things are related by descent from a common ancestor, so 'breeds' and 'species' only represent different degrees of divergence from a common ancestor. Ultimately, you and your breakfast are related, even if you are a strict vegetarian because plants are our distant relatives too. All species of animals, plants and microbes have evolved from common ancestors that lived at least 3.5 billion years ago. The branches in this universal family history make the tree of life, which we shall explore in chapter 2.

The tree of life traces the evolutionary relationships among species but there is also another important way in which species are connected: the web of nature. The cow that gave the milk for your breakfast had its own breakfast. Grass, the cow's main food, is so tough that it is regurgitated, to be chewed again as cud before it can be fully digested and the nutrients absorbed. The largest of the cow's four stomachs contain bacteria and other microbes that break down the grass and incorporate it into their own cells, which then pass to another stomach where they are digested into soluble substances that the cow can absorb efficiently. The grass, the cow and the creatures in its guts are all essential to the web of nature that ends in your kitchen, but they are by no means the whole picture.

OPPOSITE Breakfast, like any other meal, contains the products of many species and is a reminder of how we depend upon biodiversity for our existence.

The plants in the field where the cow grazed have their own places in the web of nature. Nitrogen is required for making proteins, DNA and other molecules essential to plants but the biggest reservoir of this element in nature is locked away in molecules that animals and plants cannot break down. The air is 80% nitrogen gas by volume but pairs of nitrogen atoms are so firmly wedded that they cannot be split to make other molecules essential to life. Only specialized bacteria have the chemical ability to break the bonds that bind nitrogen atoms together. Some of these bacteria have evolved a relationship with plants in the clover family and live in nodules on their roots, where the plant provides them with sugar and they provide it with nitrogen in a usable form. Thus equipped, clover and related plants are a major source of 'fixed' nitrogen for the entire web of nature on land. Clovers are pollinated by bees, and so the web of ecological connections spreads, weaving more and more species into the ecosystems that feed us at every meal. This web of nature, as well as the species in it, is also a manifestation of biodiversity.

Fragile web is about biodiversity: what it is, how it evolved, what is happening to it, what can be done to protect it and why it matters. There are at least four good reasons why you might or should care:

- because our survival depends upon the biodiversity that supports life on our planet
- because biodiversity is useful (the breakfast argument)
- because biodiversity is aesthetically pleasing
- because humans are directly or indirectly responsible for putting many other species in jeopardy – what *moral right* do we have *not* to care?

Persuasive though these reasons are, perhaps none of them would ultimately matter if the fate of biodiversity were sealed but, thankfully, it is not. There are many things that can be done to protect it and locally to even enhance it.

The biodiversity breakfast is a mere taster for the full story of how much we directly depend upon other organisms for our own existence. The consequence of this dependence is that our demands have an enormous impact on biodiversity. This impact is both direct, through harvesting what we need and creating new kinds of organisms by domestication and cultivation, and indirect, through the way in which we alter, pollute and destroy the habitats where other species live. The human ecological footprint generated by our demands on nature grows ever larger.

The human ecological footprint

The impact of humans on the Earth is determined by the numbers of our species, our lifestyle and affluence (because wealthy individuals use more resources) and the state of technology.

According to United Nations estimates, during the four decades from 1950 to 1990, the human population doubled from about 2.5 billion to 5 billion. In 2010, the population will be around 7 billion and by 2050 it seems certain to be near 9 billion, or well over three times as many people as there were a century earlier. What effect is our rising population having upon the natural world? One way to measure the impact is to

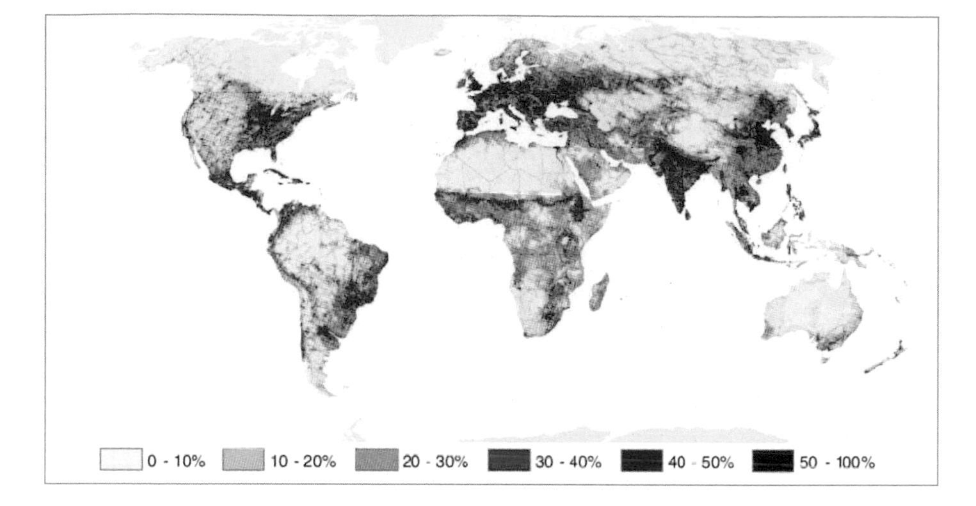

LEFT The human footprint on Earth. The human impact is expressed as the percentage of human influence relative to a maximum. Data include population density, land transformation, electrical power infrastructure and access to land.

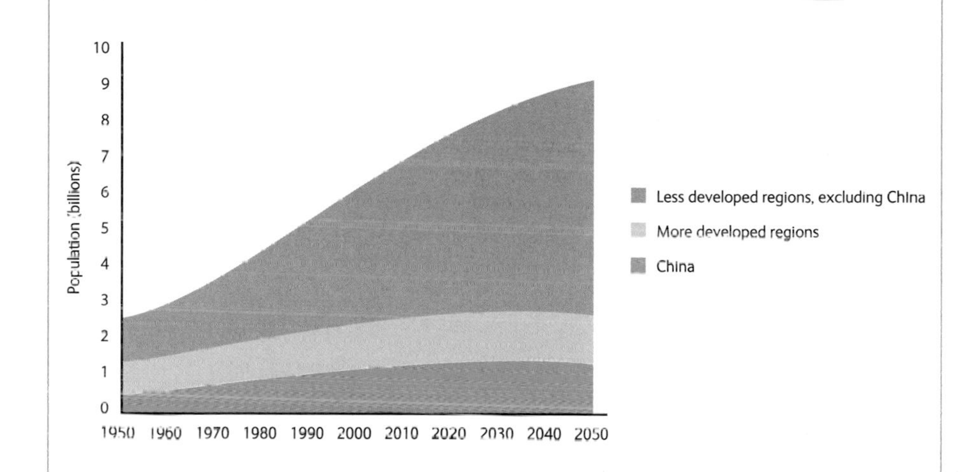

LEFT World population and its projected increase up to the year 2050 according to United Nations estimates.

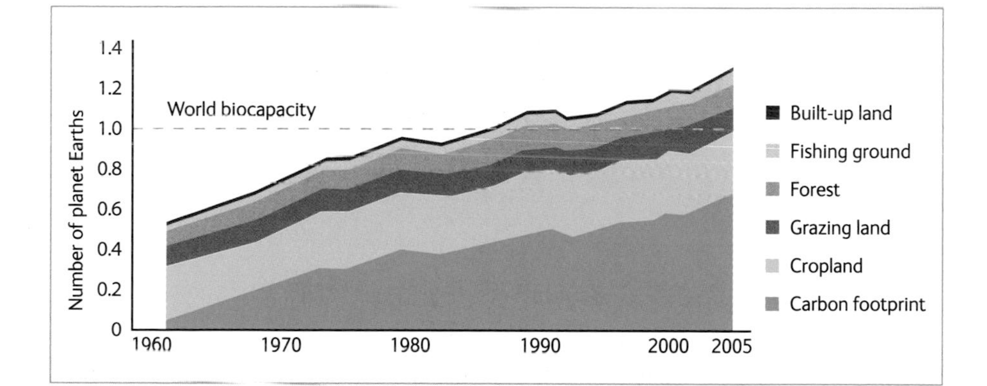

LEFT The human ecological footprint, measured as the area of the planet required to support our demand on resources and to process waste such as CO_2. Six major components of the footprint are shown, illustrating how the increase in the production of CO_2 has pushed our footprint beyond the capacity of the Earth to bear it.

translate the different resources that we use into a common currency based upon the area of the Earth required to produce them or to absorb the waste products such as carbon dioxide (CO_2). The calculations for such an exercise are full of assumptions but they show that the biggest factor in increasing humanity's ecological footprint is the huge rise in CO_2 emissions.

Carbon dioxide is produced by organisms, including ourselves, converting food to energy and by burning a carbon-based fuel such as oil, gas, coal or wood. Carbon dioxide is removed from the atmosphere by green plants and other kinds of organisms capable of using sunlight to drive chemical reactions, including teeming numbers of minute phytoplankton that live near the surface of the oceans and freshwaters. In this way the living world has the capacity to process the waste CO_2 generated by organisms, converting food to energy. However, that capacity has its limits and our activities, especially burning fuels, are now exceeding them, which is why the concentration of CO_2 in the atmosphere is increasing by about 3 parts per million every year. This amount sounds trivial but accumulating over time, it has produced a 50% increase in the concentration of CO_2 in the atmosphere since pre-industrial times.

The ecological footprint graph on p.9 shows that our ecological footprint is now bigger than the Earth can support and it has overshot our planet's capacity to sustain our activities. We are now in growing ecological debt to nature. How much of this debt is caused by the increase in human population and how much by an increasingly affluent lifestyle? One way to answer this fundamental question is to compare how population and each individual's share of the global footprint have changed over recent decades in different regions. Although the footprint per person increased in North America and Europe, globally the average footprint per person has remained constant since about 1975 while the global population has continued to rise. Increasing consumption in rich countries is certainly part of the problem but so is the sheer number of people on our overcrowded planet. Both consumption and population are part of the equation of human impact and neither can be ignored. While the human population has grown, what has happened to populations of other species? How is nature paying the price?

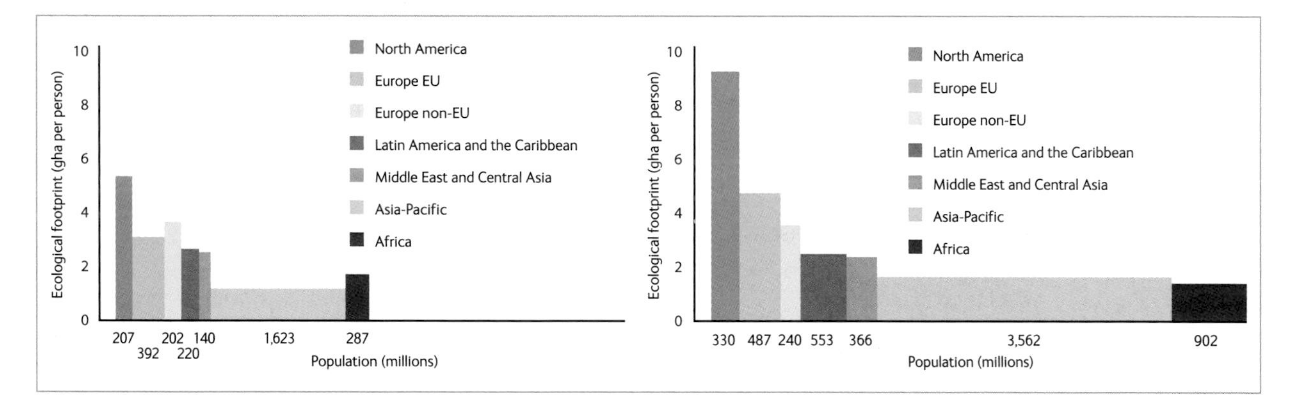

BELOW The ecological footprint per head of population for different regions of the Earth in 1961 (left) and 2005 (right). Growth of the population over the period as well as an increase in the footprint per person in the developed world, have both contributed to a large increase in the total.

Paying the price

The ecological footprint graph is so dominated by the rise in atmospheric CO_2 that it obscures how direct impacts through agriculture, fishing, deforestation and changing land use have diminished natural habitats. To measure the state of the living world, the World Wide Fund for Nature uses an index based upon estimates of the abundance of 5,000 populations of wild animals belonging to over 1,500 species around the world.

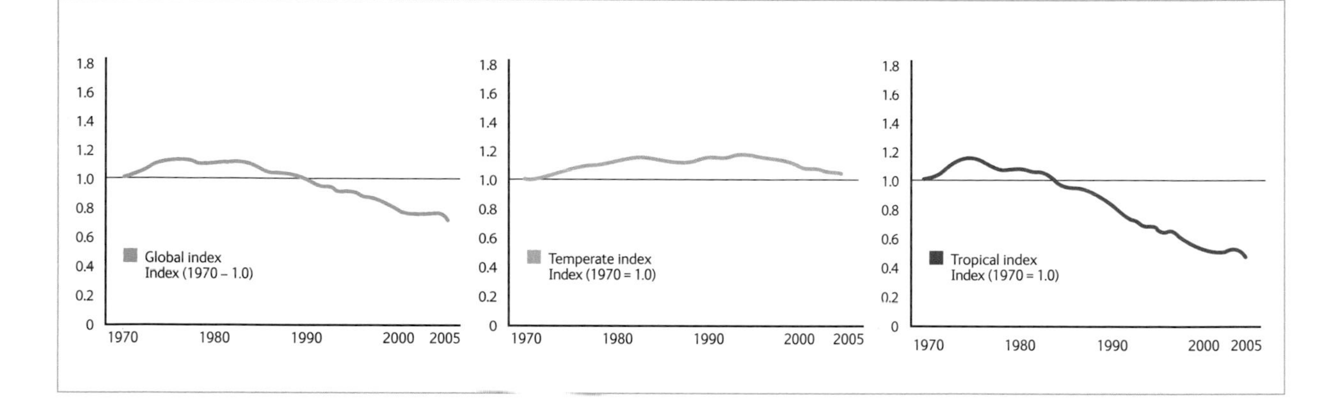

ABOVE Change in the Living Planet Index between 1970 (the base line) and 2005. The global value of the index has declined significantly over the period, mainly because of a large decline in the tropics.

This Living Planet Index is designed to track nature, rather as share price indices such as the Dow Jones or the FTSE 100 track stock markets. Starting from an arbitrarily defined baseline of one in 1970, the index has remained around this value in temperate regions but declined by 50% in the tropics where habitat loss through deforestation is a particularly severe problem.

Just as in stock markets, there are winners and losers. The Living Planet Index for animals in Western Europe has done well, increasing by 20% in the last decade, probably in response to extensive efforts in nature conservation and pollution control. Globally, however, the index for birds has declined by 50% and for mammals by 30%. We shall look in more detail at what these figures mean in chapter 11 but numbers and indices do not really convey the full ecological meaning of losing biodiversity: full understanding needs an object lesson.

A warning from Easter Island

At sunrise on 13 June 1772, Captain James Cook weighed anchor and, chased by a brisk gale, sailed out of Plymouth harbour with his ships *Resolution* and *Adventure* on his second voyage to the southern Pacific Ocean. His mission was to search for an undiscovered continent that he believed must lie farther south than any ship had yet sailed. Two years later, after the expedition had been the first to cross the Antarctic Circle and had spent months in polar waters, Cook and his men were seriously ill with scurvy, a disease caused by a deficiency of vitamin C in the diet. The Scottish surgeon James Lind had recently demonstrated that citrus juice and vegetables could prevent and cure the disease but before they left Plymouth, another doctor had claimed that malt was more effective and by taking this erroneous advice, the expedition had neglected to carry effective remedies for scurvy. Cook himself may have only survived because the ship's doctor served him broth and fresh meat from the last animal alive on board – the doctor's own pet dog. In desperate condition, Cook navigated northwards and on 11 March 1774, 103 days after their last view of land, they sighted Easter Island, the most remote inhabited island on the planet. The joy

ABOVE Captain James Cook arrived at Easter Island in 1774 and stayed for four days.

with which Captain and crew must have heard the lookout cry 'Land ahoy!' can only be imagined but what lay ahead was no island paradise.

What they saw as they drew nearer was an almost barren, volcanic islet guarded by cliffs of black basalt. Despite living over 3,219 km (2,000 miles) from the nearest mainland, the Polynesians who inhabited the island had seen European visitors before. The islanders were friendly and traded sweet potatoes, bananas, sugar cane and chickens but Captain Cook stayed just four days and later wrote:

> '*No Nation will ever contend for the honour of the discovery of Easter Island as there is hardly an island in this sea which affords less refreshments and conveniences for shipping than it does.*'

Remarkably, this barren place almost devoid of trees boasted hundreds of huge basalt statues, some crowned with enormous caps of red stone. The statues had been hewn from quarries and transported around the island where they were erected in groups facing inwards towards the barren landscape. Like every visitor since, Cook marvelled at the feats involved in their creation:

> '*We could hardly conceive how these islanders, wholly unacquainted with any mechanical power, could raise such stupendous figures, and afterwards place the large cylindrical stones upon their heads.*'

Inevitably, in the twentieth century, some speculated that the sculptures were the work of bored extraterrestrials stranded on the island by a crashed spacecraft, but reconstruction of the environmental history of the island using archaeological evidence and plant and animal remains tells a much more sobering story. Plant remains, particularly pollen buried in peat in one of the main volcanic craters on the island, show that there were once large trees growing there and that the whole island may have been covered in rainforest when the Polynesians first arrived, perhaps as recently as AD 900. The largest tree was a palm related to the huge Chilean wine palm. The biggest specimens growing on Easter Island had trunks over 210 cm (7 ft) in diameter. This species and more than 20 other plants including many trees are now extinct there. What happened to them? We shall never know exactly, but extensive archaeological evidence strongly supports the theory that Easter Island experienced an ecological collapse caused by the over-exploitation, and ultimately the total destruction, of its forest.

When Captain Cook's expedition arrived, it is estimated that the population of Easter Island was about 4,000 people but in earlier centuries it had been much, much larger. The foods that the islanders traded with the Cook expedition are found in every Polynesian culture and must have been brought to the island by the first settlers. The early forest would have been cleared to grow crops and to raise chickens, which were kept in long, purpose-built stone houses. The remains of over a thousand chicken houses still dominate the coastal areas of the island.

For the first few generations after Polynesian settlement, Easter Island would have been a gourmet paradise with plentiful vegetables, chicken, locally caught fish, seabirds, porpoise, turtles and the edible Polynesian rat that the settlers had also brought with them. With food so plentiful, the population grew and there was labour to spare for

ABOVE The remains of over a thousand chicken houses on Easter Island stand testament to the large human population that the island once supported.

more than just subsistence. Statues were carved and erected for ancestor worship and it appears that rival clans competed with one another in the stature of their statuary, carving bigger and bigger figures. The very biggest ones still lie uncompleted in a quarry on the Island, including the largest that is 21 m (70 ft) long and weighs 270 tons. It could never have been moved but lies there, testament to a mania for ostentation that can be recognized among the wealthy and powerful in all human societies. As the human population of Easter Island grew, more and more forest was cleared. Trees were also felled for fuel and for logs that were used to build canoes and probably to create wooden tracks to transport the statues along. When, in the 1950s, someone finally thought to ask the islanders themselves how their ancestors erected the statues, they demonstrated the technique by building ramps and using logs as levers.

The charcoal remains found in fireplaces documented the extinction of the forest and showed that by about 1640, even the elite of island society were burning only grass and herbs in their hearths. With the trees all gone, life on Easter Island changed totally. Seaworthy canoes could no longer be made, which put important wild foods such as oceanic fish, turtles and porpoise out of reach. Islanders turned to birds, driving all the land species and two-thirds of the seabirds to extinction. With no tree roots to hold it, massive erosion washed the soil from the hills and the productivity of the inland plantations fell. Before statue production ceased altogether, smaller carvings were made that graphically represent starving people with hollow cheeks and jutting ribs. Even the dead were affected by the crisis because there was insufficient fuel for

ABOVE The famous statues of Easter Island were erected when the population prospered. Environmental degradation led to the collapse of society and warring groups overturned each other's monuments. The statues shown here were re-erected in modern times.

traditional cremation. Civil war broke out and clans began toppling their rivals' statues. Some were still standing when Cook visited in 1774 but by the middle of the next century, all had been deliberately felled.

Ecological collapse on Easter Island is an obvious metaphor for where the loss of biodiversity may lead the entire planet, though it would be simplistic to expect global events to follow exactly the same course as happened on a small oceanic island. Nevertheless, there are at least four lessons from Easter Island that we ought to heed:

- Living natural resources such as forests have a limit, called their *carrying capacity*, beyond which over-exploitation and collapse will occur.
- Removing a part of the web of nature, palms in the case of Easter Island, can lead to the degradation of the whole web.
- Collapsing societies are not uncommon in human history and the event is often triggered by an over-exploitation of natural resources.
- Islands and other isolated environments are especially vulnerable to over-exploitation.

Some islands are more vulnerable than others. A survey of deforestation on more than 80 islands found that the most badly affected ones were those that are small, remote, outside the tropics, have a dry climate and have infertile soil. This pattern suggests that islands with better conditions for plant growth (warm, wet, fertile) have more resilient forests: if cut down, they can regenerate more quickly and completely. Small areas are more easily cleared; remoteness offers the inhabitants no chance of escape to new places and consequently their island has no opportunity for respite from exploitation. Easter Island had the misfortune to be habitable when first colonized, and to suffer from nearly all of the environmental factors that make an island vulnerable to deforestation. We shall have occasion to remember the lessons of Easter Island in later chapters but first we shall put the little local difficulties on that remote island into perspective by surveying the broad sweep of life's history on Earth and the origins of biodiversity.

CHAPTER

2 *Life's long and chequered career*

FOR MOST OF ITS HISTORY THE EARTH has been inhabited by life. Geological clues include both the fossilized remains of ancient organisms and evidence for the conditions in which they lived. Given the vast span of time since the Earth's formation some 4,600 million years ago and the sporadic deposition of sediments in which such evidence might be preserved, the record is understandably patchy. Nevertheless, geologists have been able to piece together a timescale of successive *eons*, the last of which (the Phanerozoic) is subdivided into *eras* and they in turn into *periods*.

OPPOSITE Up to a metre long, the Cambrian marine predator, *Anomalocaris*, pursues a prospective victim. The evolution of offensive and defensive hard parts in response to escalating predation greatly enriched the Phanerozoic fossil record.

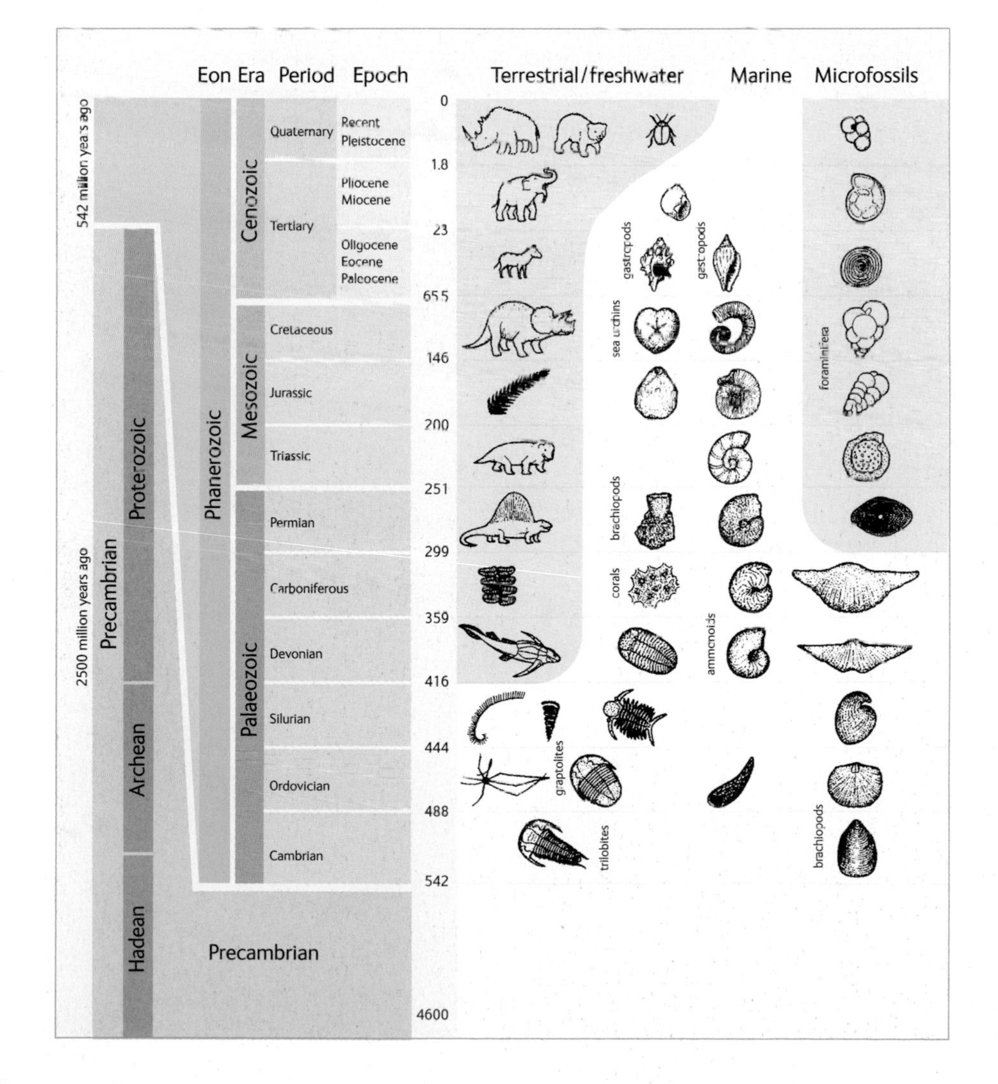

LEFT The geological time scale. The Phanerozoic Eon (green) is expanded to show its constituent eras and their subdivisions, with some typical fossils, to the right. Numbers indicate millions of years before the present.

Planet of the microbes

The story of life starts with the unimaginably long reign of single-celled microbes that lasted for at least three-quarters of life's history on Earth. Fossils of simple early life forms are unsurprisingly scarce and difficult to interpret – even to distinguish from the products of non-living processes in some cases. Chemicals that, as far as we know, are produced only by living organisms nevertheless provide some further clues.

For want of direct evidence, ideas about how life began rely upon little more than informed guesswork. However it began, we can assume that sooner or later discrete cells emerged, enclosed by organic membranes that mediated chemical exchange with their surroundings, as in the cells of all life on Earth today. By processing materials acquired thereby (metabolism) and using energy derived in one way or another from their environment, they could grow and maintain their integrity in changeable conditions. They could also reproduce themselves with high fidelity thanks to their contained genetic material, as we can infer from its universal presence in today's life forms. These twin capabilities – metabolism and reproduction – were the essential ingredients for evolution.

Archean rocks have yielded a tantalizingly diverse crop of possible clues to the existence of early life, although much remains subject to debate. Nevertheless, most experts agree that microbial life forms probably became established on Earth some time between 3,800 and 3,000 million years ago, after a period of massive bombardment by asteroids, if not before then.

Two genetically distinct domains of microbes soon emerged – the Bacteria (also called Eubacteria) and the Archaea. The former include both disease-causing and free-living types found in most surface environments, and are the most important agents of decay. The Archaea, by contrast, are a disparate assemblage of forms some of which occupy various extreme (at least to us) environments, including hot springs, highly

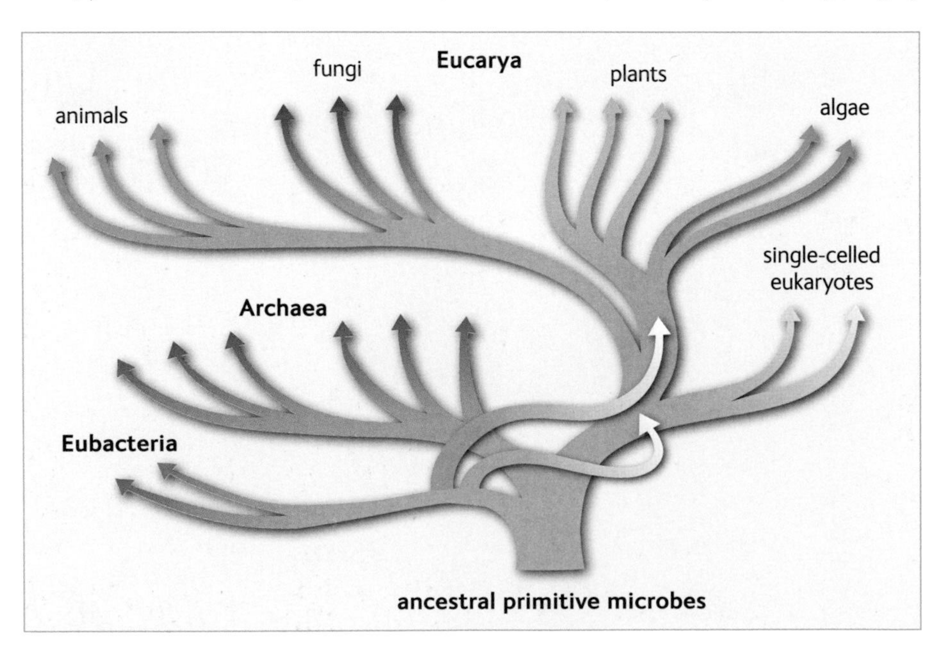

RIGHT The modern DNA-based tree of life. The branching pattern indicates evolutionary relationships inferred from comparisons of DNA sequences. Three major branches of the tree, Eubacteria, Archaea and Eucarya are indicated, as well as the main subdivisions of the Eucarya, including the three multicellular kingdoms that dominate macroscopic life. The two lines that run from the Eubacteria to the Eucarya represent the incorporation of precursors to chloroplasts (in plants) and mitochondria (in Eucarya), respectively.

salty and/or acidic habitats and settings devoid of oxygen such as stagnant mud. Such conditions were probably commonplace at the early Earth's surface, so the Archaea have attracted great interest as models of, at the least, if not actually little-changed relatives of some of the earliest life forms on Earth (their name, like that for the eon, comes from the classical Greek for 'ancient'). A third domain, the eukaryotes (or Eucarya), to which we and all other non-microbial life belong, has a curiously mixed genetic make-up, suggesting that they may have originated from an early fusion between archaean and bacterial ancestors. Eukaryote cells are generally much larger and more complex than those of the two microbial domains, and differ from them in housing most of their genes in a membrane bound body called the nucleus.

The beginning of solar power

One important group of Bacteria that had evidently arisen by early Proterozoic times were the cyanobacteria. Today, filamentous cyanobacteria (together with other microbes) form mats on sediment surfaces, where they trap the energy of sunlight and use it to convert water (H_2O) and carbon dioxide (CO_2) to sugars (from which more complex molecules are assembled) in the process known as photosynthesis, releasing free oxygen (O_2) as they do so. As they grow, these mats also trap fine particles of

BELOW LEFT Large fossil stromatolites of Late Proterozoic age in a natural outcrop in Oman (cm scale in foreground).

BELOW TOP Wrinkled microbial mat forming similar structures to fossil stromatolites, today, in the intertidal sabkha environment of Abu Dhabi in the Arabian Gulf.

BELOW BOTTOM Thin section through a 2,000-million-year-old stromatolite from Canada, showing fossilized microbes (cyanobacteria). The spheres are about 0.01 mm across.

lime sediment washed over them, like dust collecting in the pile of a carpet, so creating successive thin layers of trapped sediment that can build up to form lumpy-looking banded deposits known as stromatolites. Fossil stromatolites, are abundant in Proterozoic rocks and their cyanobacterial origin is confirmed in some cases by the petrified remains of cells. Chemical traces of likely cyanobacterial origin date back to the Late Archean (about 2,800 million years ago), and structures resembling stromatolites have even been found in rocks as old as 3,500 million years in Australia, although whether these latter examples were really of microbial origin or inorganically produced is still debated. We have to imagine, then, that maybe for the first billion years (1,000 million years) or so of life's history, Archean to Early Proterozoic seascapes would have presented a monotonous scene of slimy microbial mounds flanking barren continental hinterlands.

Nevertheless, one crucial long-term effect of such photosynthesizing microbes was to endow the atmosphere with free oxygen, thereby radically altering the conditions for subsequent evolution. Initially, until well into Early Proterozoic times, most of the oxygen produced was consumed by reaction with large quantities of ferrous (Fe^{2+}) ions contained in hot brines emanating from fissures in the ocean floor linked with submarine volcanic activity. Evidence for this massive 'mopping up' of oxygen comes from contemporaneous 'banded iron formations', which contain vast quantities of oxidized iron (ferric or Fe^{3+}-bearing oxides).

RIGHT Cut face about 1 m (3¼ ft) high on a block of 2,250-million-year-old banded iron formation, from Michigan, USA. In this example, layers of red jasper alternate with grey iron oxide.

ABOVE Middle Proterozoic red beds exposed in a cliff in Greenland, reveal the point in geological time when free oxygen built up in the atmosphere.

By about 2,000 million years ago, the balance had swung in favour of the steady build-up of free oxygen in the atmosphere (with concomitant reduction of CO_2), as is shown by the appearance in the geological record of 'red beds' – sedimentary deposits formed on land and deeply pigmented by red iron oxide (haematite, Fe_2O_3) as a result of weathering in oxidizing conditions.

Although we depend on oxygen ourselves, this new component of the water or air around them was poisonous to most of the pre-existing microbial forms of life. Many became adapted to environmental refuges that remained unaffected by the increase in oxygen. That is perhaps why archaeans are abundant in what we regard as 'extreme' habitats, including hundreds of metres deep within ocean-floor sediments (the so-called 'deep biosphere') – such places have remained somewhat like their ancestral homes. A few have even found comparatively new congenial anoxic niches in the guts of animals such as cattle, where they play an important role in methane production. Some bacteria, by contrast, became adapted to tolerate free oxygen by using it in their metabolism, probably in association with their own photosynthesis, as seen in the living 'purple bacteria'.

It appears that some of these oxygen-processing bacteria became incorporated within the cells of ancestral eukaryotes. There they remained in symbiotic union with their hosts, eventually to become the tiny intracellular bodies, or organelles, known as mitochondria (singular mitochondrion). In living eukaryotes, the latter perform the function of reacting sugars or fats with oxygen to yield water, CO_2 and, importantly, lots of energy. This process is termed aerobic respiration. The best evidence for this theory of their origin is the close similarity between the small amounts of genetic material in the mitochondria and that found in purple bacteria. This arrangement paid adaptive dividends both ways. For the bacterial precursors of mitochondria, a stable micro-environment for proliferation, nourished by a ready supply of sugars, was provided by their eukaryote hosts. The latter in turn benefited from the greater energy yield of the aerobic respiration, compared with the incomplete chemical breakdown involved in the older style of anaerobic (non oxygen-consuming) respiration. Nor did the microbial coalitions stop there. Similar genetic evidence indicates that the photosynthesizing organelles in plants, known as chloroplasts, were likewise derived from other bacteria.

THE UNIQUE TALENT OF CYANOBACTERIA

Life on Earth is very largely solar-powered, but all the organisms capable of capturing solar energy by means of photosynthesis either belong to the cyanobacteria or they acquired this ability from cyanobacteria during the course of evolution. For example, in land plants and eukaryotic algae such as seaweeds and pondweeds, photosynthesis occurs inside organelles called chloroplasts. The light-absorbing pigments in these structures give plants and algae their green colour. Chloroplasts, like mitochondria, have their own genes and resemble bacteria, which is a clue to their origins. By comparing chloroplast genes from different species of plants it has been found that all land plants, from tiny mosses to giant redwoods, descend from the same common ancestor that made the transition from ocean to land. That common ancestor carried the forebear of all chloroplasts.

Free-living cyanobacteria still inhabit the sea and freshwater, but some of them were engulfed by eukaryotic cells, forming a permanent partnership that gave rise to two groups of algae, one green and one red (the latter possess the green pigment essential to photosynthesis, but it is masked by the red colour of another pigment). The original algae were single-celled, but their subsequent evolutionary careers diverged. While some continued a single-celled lifestyle in the sea or freshwater, others became multi-celled and one lineage of green algae colonized land, eventually diversifying into all the plants we are familiar with today. You could say that trees are green algae on wooden pedestals.

Back in the ocean and following a different route, a marine red algal cell with its internal cyanobacterial passenger was in its turn engulfed by another cell and a new, permanent partnership was formed that evolved into brown algae (think of the brown seaweeds revealed on any rocky seashore by low tide) and diatoms. Diatoms are tiny photosynthetic cells that form an important part of the phytoplankton in the oceans, capturing between them as much CO_2 in a year as rainforests do on land.

ABOVE Evolutionary tree for photosynthetic life. All land plants descend from the same common ancestor that made the transition from ocean to land.

Eukaryotes invent sex

Although the age of the oldest fossils of probable eukaryotes (about 1,900 million years old) seems conveniently close to the time of the appearance of free oxygen in the atmosphere, distinctive chemical traces ('steranes') of their cell membranes preserved in ancient rocks in western Australia suggest, rather, that eukaryotes were already around nearly 2,800 million years ago.

From the time of the rise of atmospheric oxygen, we can fast-forward another billion years or so to the next major revolution in life's history – the evolution of sex. The essence of sex is that it brings together genes from a pair of individuals, the parents, in one generation to form a new combination in their offspring in the next generation. However this bizarre process came about (a subject of much debate), it had a profound impact on evolution. Instead of producing genetically near-identical clones as in bacterial-style asexual reproduction, the shuffling of parental genes involved in sexual reproduction yields genetically diverse offspring. As a result, natural selection can operate with much greater efficiency on virtually every conceivable combination of genes in each generation and screen out disadvantageous ones, allowing relatively rapid adaptive responses to differing circumstances. A further consequence was the new-found scope for splitting of populations into distinct species whenever barriers to reproduction arose between them, with divergence following on as each then followed its own evolutionary pathway.

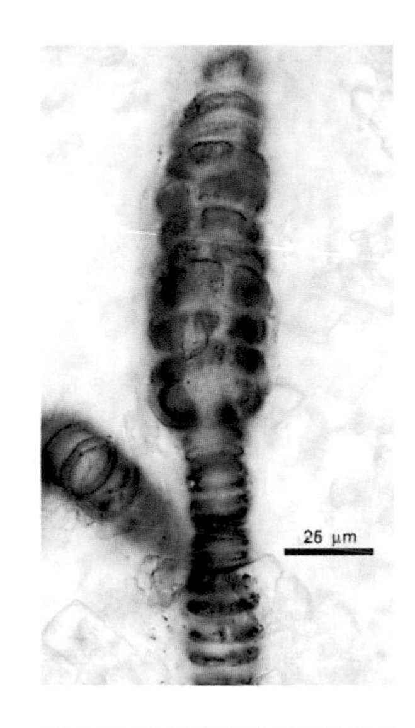

ABOVE The termination of a *Bangiomorpha* filament containing large numbers of probable spores.

We have no direct record of how this reproductive revolution began, but comparisons of the genes of living eukaryotes, supplemented by fossil evidence suggest that it occurred around 1,000–1,200 million years ago. From then onwards, the fossil record testifies to an increasing proliferation of eukaryote life forms, both unicellular and multicellular, with the oldest supposed multicellular alga, *Bangiomorpha*, being recorded from 1,200-million-year-old strata in arctic Canada. The appearance in these fossils of distinct gametes, moreover, makes them the oldest known example of sexual reproduction. This 'Big Bang' of eukaryote evolution gave rise to a plethora of single-celled groups as well as to the familiar multicellular kingdoms of plants, fungi and animals.

Complex food webs appear

Despite having now covered some 3 billion years of life's history, this discussion has been virtually limited to estimating when and, in the most general terms, how successive grades of complexity were attained, from the origin of life, to microbes, to eukaryotes. The diversification of these groups has scarcely been touched upon because of the paucity of reliable evidence. There are not enough rocks preserved in suitable condition to chart the changing map of the world through this time, nor, therefore, the geographical diversity of life.

However, towards the end of the Proterozoic, the record starts to become more informative, just in time to drop some intriguing hints about the circumstances in which the first animals evolved. Between 850 and 630 million years ago the Earth experienced

RIGHT Glacial deposits of Late Proterozoic age (650 million years old) in Ella Ø Island, Greenland.

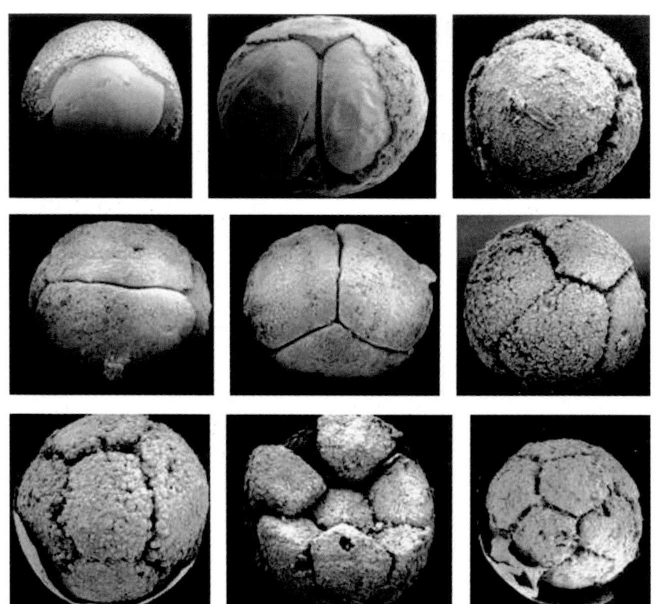

ABOVE Exceptional fossils of animal embryos, preserved as phosphatized replicas of the original cells, from the approximately 600-million-year-old Doushantuo Formation of China.

two successive ice ages, which were probably the most intense in its entire history. Arguments persist about whether ice extended at sea level into the tropics (the 'snowball Earth' theory), but the presence of diagnostic glacial deposits on several continents indicates very widespread glaciation.

Some authorities postulate a link between the amelioration of conditions at the end of the second major episode of glaciation, 630 million years ago, and the rise of animals. The coincidence of less than pinhead-sized fossils of this age, interpreted as likely animal egg cases, is at least suggestive of such a link, although it is unclear what the causal connection might have been. A little younger than the latter (maybe about 600 million years old) are exceptionally well-fossilized (phosphatized) balls of cells found in the Doushantuo Formation in southwestern China that are widely agreed to represent animal embryos.

From an ecological perspective, the importance of the appearance of animals is that it opened the way for the development of complex food webs. Whereas plants produce their own food by photosynthesis (*see* p.22, The unique talent of cyanobacteria) and fungi absorb the molecular breakdown products of other organisms, animals feed directly on other organisms and/or their remains. Although the latter habit is not exclusive to animals, as many single-celled eukaryotes do likewise, the vast range of sizes and forms attainable through multicellular development in animals allowed hierarchies of eaters and eaten – or trophic pyramids – to build up, as in the proverb 'big fish eat little fish'.

LEFT Reconstruction of the shallow seafloor with Ediacaran animals, about 570 million years ago.

From the time of the Doushantuo fossils to the beginning of the Phanerozoic Eon, a radiation of enigmatic seafloor-dwelling animals ensued, represented by fossils found on several continents. Named after a famous site in southern Australia, the Ediacara Hills – where a diverse assemblage is preserved as impressions in beds of sandstone – this 'Ediacaran' fauna was entirely soft-bodied, that is to say, lacking skeletal components. Although some reached a metre or so in length (about 40 in), they were relatively thin, possibly as an adaptation to direct absorption of still relatively low levels of oxygen over their external surfaces. It remains a matter of debate whether they were early representatives of extant groups such as cnidarians (e.g. jellyfish and sea anemones) and flatworms, or a distinct grouping of animals most of which became extinct at the close of the Proterozoic.

With the dawn of the Phanerozoic Eon, however, a dramatically new picture of animal life is presented. A rapid proliferation of resistant skeletal hard parts led to a massive enrichment of the fossil record – hence the name of the eon, which means 'visible life' in classical Greek. Deposits even of latest Proterozoic age, by contrast, yield only a paltry variety of minute conical shells besides the fossils of the soft-bodied Ediacaran forms mentioned above. Numerous explanations have been offered for this change. Although some connection with rising levels of oxygen has long been a favoured hypothesis, a combination of evidence from comparing the genes of extant groups and from fossils suggests that innovations within the organisms in question may have played an important part as well.

The fossils themselves hint strongly at one factor that may have stimulated the evolutionary proliferation of animal body plans – escalating predation. Even some of the tiny tubular shells of the latest Proterozoic age show circular drill-holes suggestive of predatory attack on their occupants, while clear examples of predators, predation damage (with subsequent healing) and defensive coverings of plates and spines (like chain-mail) can all be seen in Cambrian fossils: 'nature red in tooth and claw' was here to stay.

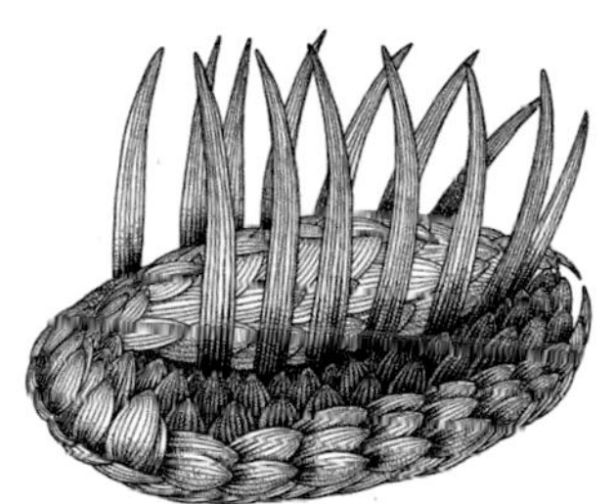

ABOVE Signs of predation in Cambrian animals: (above) reconstruction of an armoured worm, *Wiwaxia*, from western Canada; (below) *Modocia*, a trilobite with a healed bite mark on the left side, from Utah, USA.

Life's ups and downs in the Phanerozoic

BELOW Changes in diversity of marine invertebrate genera recorded through the Phanerozoic fossil record.

Thanks to these changes, the Phanerozoic record provides a markedly improved picture of life's ups and downs up to the present day. That of marine invertebrates gives the most accurate account, because of the relative abundance of robust skeletal remains (especially shells) and the likelihood of these remains being entrapped by sediment eroded from the land and washed into the sea.

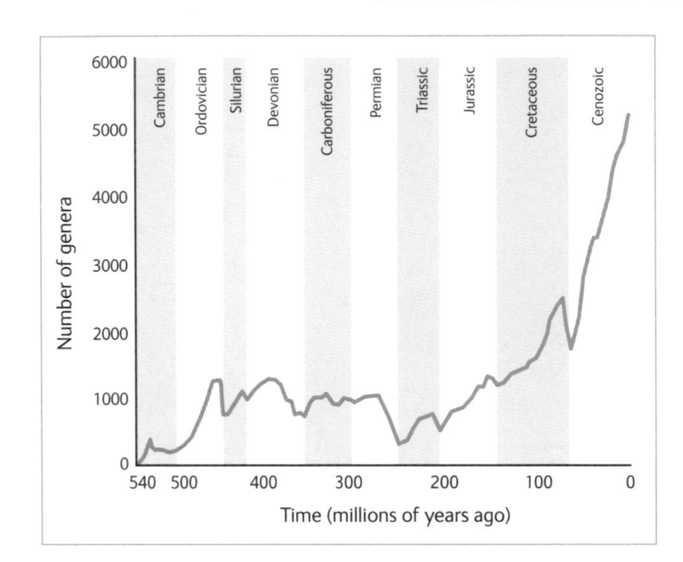

Two main factors contributed to an overall increase in biodiversity during the Phanerozoic, albeit repeatedly set back by mass extinctions of varying intensity. First, the expansion of life onto the land added whole new ecosystems with their myriad denizens to the planet. Second, episodic escalation of the scope and intensity of ecological interactions, like that of predation at the dawn of the eon, further supplemented the variety of living organisms, both on land and in the sea.

Among marine organisms, for example, during the long Mesozoic recovery of life from the devastating mass extinction that occurred at the close of the Permian (marked by a sharp drop in the number of genera in the graph), there was a further radical escalation of interactions between predators and their prey. New groups of specialized predators arose, including various kinds of crabs, starfish and snails, for example, who respectively smash, wrench open and drill through the shells of their prey. Under this onslaught, the prey in turn evolved a wide repertoire of defensive adaptations, including stoutly buttressed or spiny shells, deeper burrowing (tens of centimetres) into sediment than their Palaeozoic forebears had ever done, boring into stiff mud or rock, cementing to hard surfaces (as in oysters) or simply casting off from the bottom to escape.

RIGHT Some silent seashore predation as a dogwhelk inserts its proboscis into one of the smaller mussels attached to the rocks, and gorges on its soft parts.

Others became adapted to clustering in physically stressful 'refuges' from many (though not all) of their pursuers, such as high on the intertidal zones of exposed shorelines. The ongoing struggles between predators and prey that can be witnessed along any coast today are a legacy of this major Mesozoic re-structuring of marine food webs.

Turning to the land, although microbial communities probably occupied marginal environments even in Precambrian times, the fossil record suggests that it was not until the Late Ordovician that the first eukaryote invasions followed on, in the form of ground-hugging, moss-like plants derived from green algae. Not until earliest Devonian times did upright-growing bushy plants (some tens of centimetres high), supported and sustained by tough internal vessels, begin to form thickets on coastal and river flood plains. Yet by the end of that period, the land vegetation already included tiered forests, with leaf-bearing trees competing with one another for access to light.

Where the plants advanced, animals followed – at least those whose body plans allowed adaptation to the tough physical challenges of living on land, especially the loss of buoyant support from water and the risk of desiccation. In fact, few of the more than 30 major groups (phyla) of animals have successfully invaded the land. Yet two such groups spectacularly flourished there, hugely enhancing the Earth's biodiversity – first, the arthropods, with their tough external cuticles (including the myriads of insects, as well as spiders, mites, scorpions, millipedes and centipedes), and second, the vertebrates. From earliest Devonian times, herbivory was the almost exclusive preserve of arthropods, with the first land-dwelling vertebrates of the Late Devonian merely topping trophic pyramids as consumers. In the Permian, however, some vertebrates began to switch to herbivory, heralding the formation of terrestrial food-webs more like those of today.

With the establishment of such communities, Late Palaeozoic landscapes became increasingly mantled by rich soils, where in previous periods there would have been only deserts of loose sedimentary debris sparsely populated by microbes. So vast did this new accumulation of composted organic debris become that it impacted in turn on the composition of the atmosphere, as the spread of cyanobacteria in Early Proterozoic oceans had done some 2 billion years earlier. The photosynthetic conversion of CO_2 into organic material stored in the land vegetation and associated soils boosted the level of atmospheric oxygen, perhaps even a bit beyond today's 21%, at the same time as reducing that of gaseous CO_2. The latter effect in turn influenced climate, by lowering the CO_2-induced greenhouse effect, so contributing to an ice-age that extended through the Late Carboniferous and Early Permian. Geological witnesses to these extraordinary events are the vast deposits of coal of this age, formed by the burial of tropical peat, and widespread glacial deposits in the southern continents. Never let it be said that we humans invented atmospheric modification!

From time to time, catastrophic environmental perturbations of global scale checked the growth of biodiversity by sweeping away large numbers of species in mass extinctions. These show up in the graph of generic diversity at the start of this section as sharp declines, most notably in the Late Ordovician, the Late Devonian and at the ends of the Permian, Triassic and Cretaceous periods (together known as the 'big five'), although there were many more, on smaller scales. Though attributable to a variety of (still debated) causes, the one thing that they all show in common is evidence for rapid

RIGHT Diorama of a Carboniferous coal swamp flanking a river. The burial of vast accumulations of peat in such tropical environments, eventually forming coal, drew down CO_2 from the atmosphere, possibly helping to provoke a major ice age over 300 million years before the Quaternary one of our own prehistory.

BELOW Diorama of a Late Cretaceous landscape in North America, with herds of grazing horned and 'duck-billed' dinosaurs, some of which are being menaced by a predator on the plain below. A mass extinction was soon to sweep them away, leaving the stage clear for the evolution of mammalian ecological replacements.

and widespread environmental change, which presumably went too far and too fast for many organisms to adapt to, leading to cascading ecological collapse.

The best understood of the 'big five' mass extinctions is the one that ended the Cretaceous period and was probably caused by an asteroid impact on the coast of the Yucatan peninsula in Mexico, where traces of a huge crater have been detected. This is the event that famously brought about the demise of the dinosaurs (or, more precisely, the 'non-avian dinosaurs', as it is now widely accepted that birds evolved from dinosaurs during the Jurassic). Other victims included pterosaurs on land and, in the sea, ammonites, belemnites and many bottom-dwelling invertebrates as well as a large proportion of the plankton. Sedimentary successions representing this time period across the world furnish abundant chemical and physical evidence for the outfall from a gigantic asteroid impact, including microscopic droplets from vaporized rock.

Not only did such catastrophes sharply reduce biodiversity, but they also cleared the way for the subsequent diversification of surviving groups that may previously have lived in the shadows, so to speak, of incumbent dominant forms. A classic case is that of the wide-ranging radiation of mammals – from bats in the air to whales in the sea – that followed the demise of the non-avian dinosaurs and other major reptilian groups of the Mesozoic Era. This is just one of the events that turned the pages of evolution's ever-changing atlas.

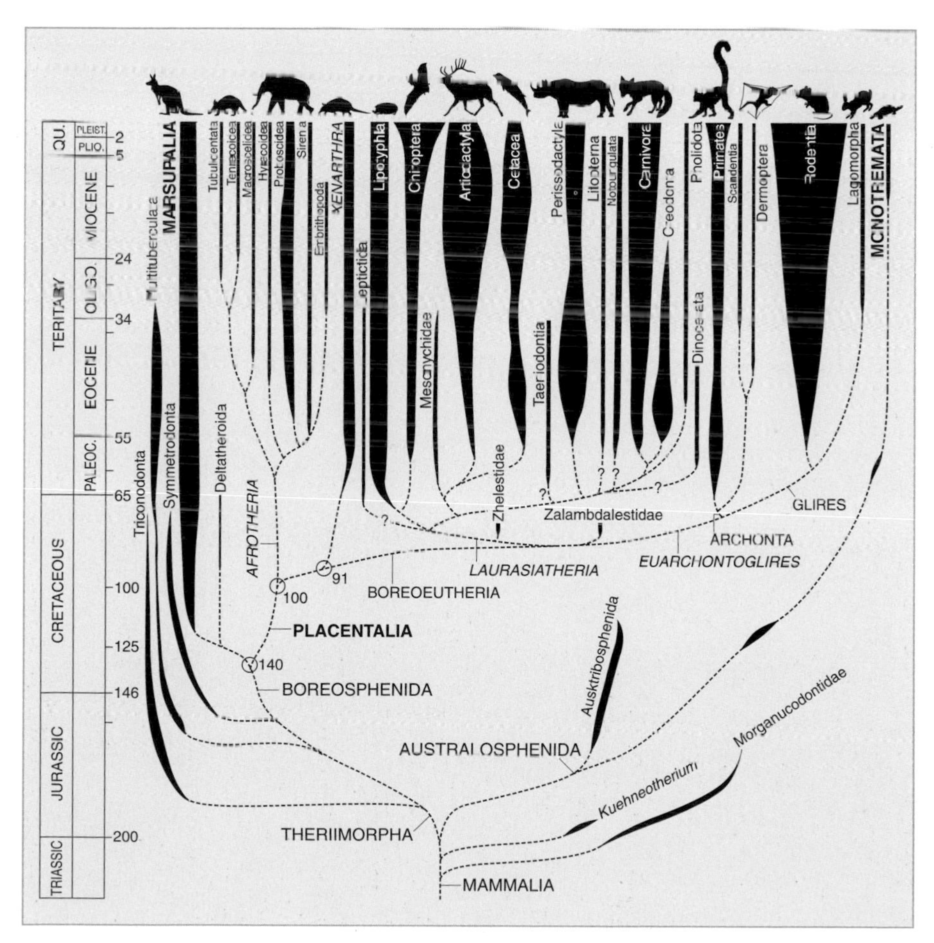

LEFT The evolutionary radiation of mammals. Branch widths show approximate diversities of the resulting groups, examples of which are illustrated by silhouettes at the top of the diagram. Relationships are inferred from comparisons of DNA, combined with fossil information, and some estimated branching dates (in millions of years) are shown. Though all the deep divisions within the tree preceded the end of the Cretaceous, overall diversity increased greatly.

3 CHAPTER *Evolution's atlas*

IN 1858 THE NATURALIST AND COLLECTOR Alfred Russel Wallace was travelling among the islands of Southeast Asia in search of specimens that he might ship home to his wealthy clients in Europe, when he was forced to detour from his intended route and so by chance visited two islands that were new to him. He later wrote in his book *Island Life* (1880):

> *'In the Malay Archipelago there are two islands named Bali and Lombok, each about as large as Corsica, and separated by a strait only fifteen miles wide at its narrowest part. Yet these islands differ far more from each other in their birds and quadrupeds than do England and Japan. The birds of the one are extremely unlike those of the other, the difference being such as to strike even the most ordinary observer. Bali has red and green woodpeckers, barbets, weaver-birds and black-and-white magpie-robins, none of which are found on Lombok, where, however, we find screaming cockatoos and friar birds, and the strange mound-building megapodes, which are all equally unknown in Bali.'*

Wallace goes on to detail other examples of this phenomenon, where a difference in faunas is so sharp that a line can be drawn between them on a map, even though no obvious physical difference between the two areas exists. Today, maps of Southeast Asian biodiversity show just such a line and it is labelled with Wallace's name. Wallace had stumbled upon a region where two faunas with quite different evolutionary histories – the Asian and Australasian – had been rafted together by geological events. It took nearly a century for those events, which have so influenced the maps in evolution's atlas, to be understood.

OPPOSITE A shrub in the protea family, one of many edemics that make the Cape Floristic Region of South Africa one of the richest hotspots of biodiversity on the planet.

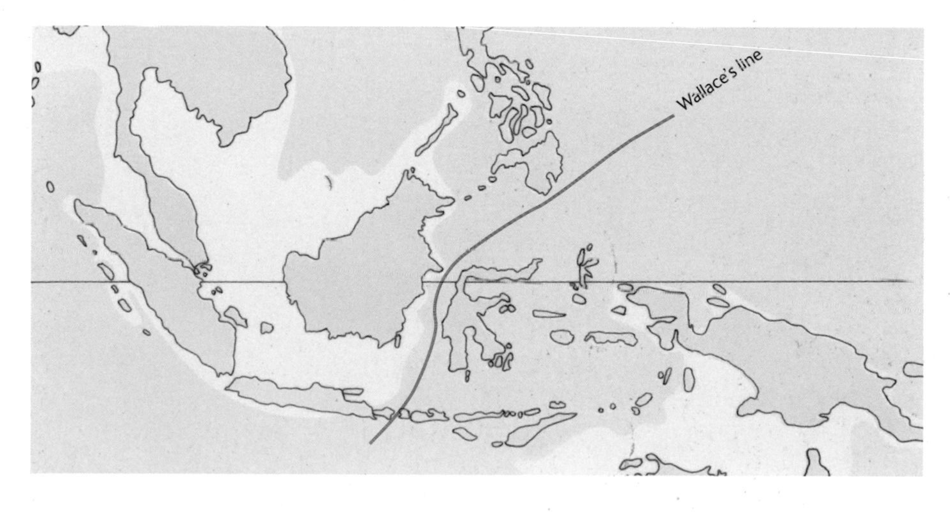

LEFT The boundary between the two very different faunas of Asia and Australasia as recognized by Wallace.

The restless Earth

The arrangement of the Earth's continents and oceans has changed continuously through time, with continents sometimes coalescing and at other times splitting apart, with important consequences for the distributions of organisms both on land and in the sea, and hence for global biodiversity.

For the shaping of today's geography, we need to go back around 250 million years, to the start of the Mesozoic Era. At this time the continental masses coalesced to form a single supercontinent, called 'Pangaea', with a large wedge-shaped ocean ('Tethys') inserting on its eastern side. Pangaea started to break up in the Early Jurassic with rifting between North America and the conjoined South America plus West Africa, to form the incipient Central Atlantic. By Late Jurassic times, opening of the Central Atlantic had extended eastwards to join up with the Tethys, so creating a seaway between what were now two discrete supercontinents, 'Laurasia' in the north and 'Gondwana' in the south.

Gondwana began to break up in the Early Cretaceous with the opening of the South Atlantic between South America and southern Africa. Later in the period, the nascent ocean joined the Central Atlantic, while the North Atlantic also began to open between North America and Western Europe. Hence, by the start of Late Cretaceous times (about 95 million years ago) there existed a continuous, though locally constricted, Atlantic Ocean. The drift of India and Madagascar from an original position nestled between Africa and Antarctica likewise gave rise to the Indian Ocean during the Cretaceous.

As the Atlantic continued to widen through the Late Cretaceous and Tertiary, Africa, with the conjoined Arabia, first rotated eastward, pivoting around Iberia, and then drifted northward relative to Eurasia, leading to the progressive closure of the western Tethys. The resulting crush of continental promontories and micro-continental

BELOW The changing pattern of Earth's continental plates through geological time. The blue indicates a period when the world was in an 'icehouse' phase, and the red when the world was in a 'greenhouse' phase.

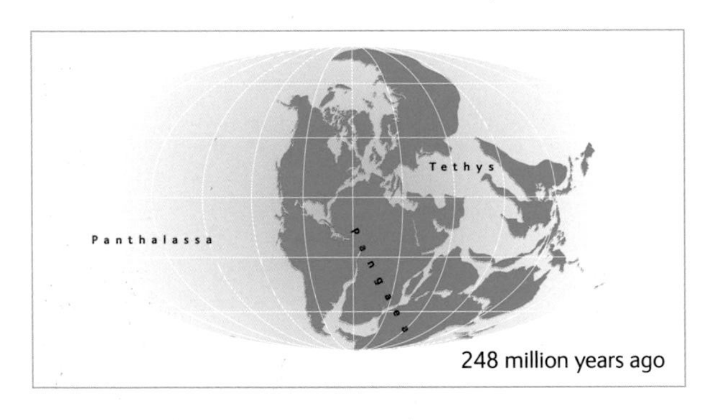

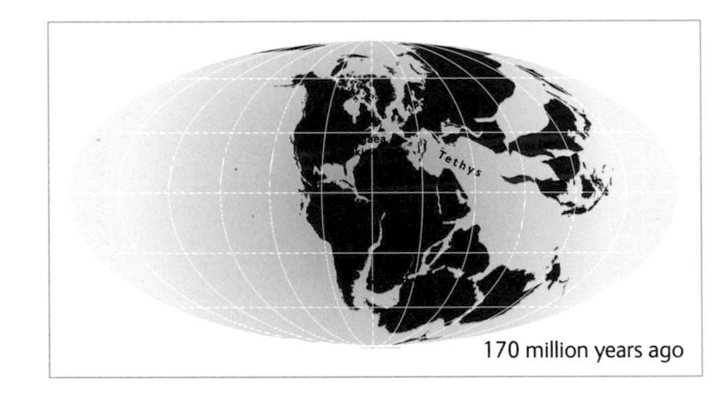

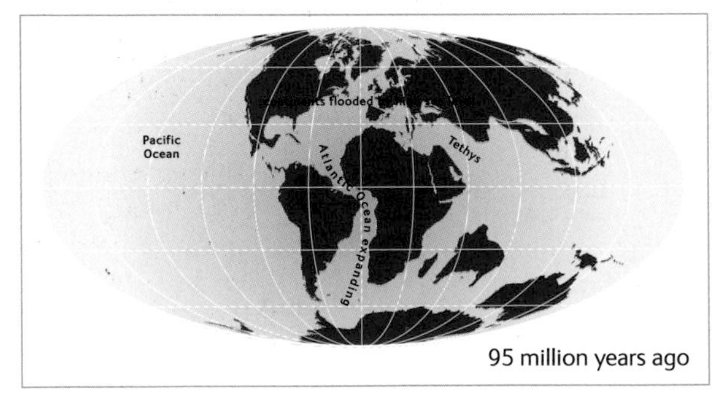

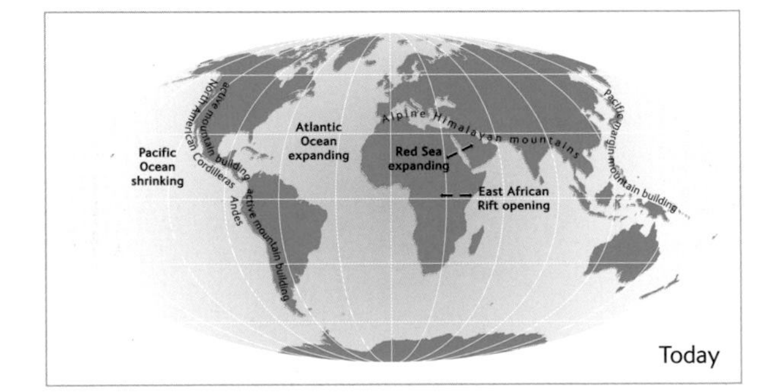

fragments between the jaw-like masses of Africa/Arabia and Eurasia created mountain chains extending from the Pyrenees to the Zagros Mountains of western Iran.

The first effects of the collisions were felt towards the end of the Cretaceous, but the main period of mountain-building ensued in the Tertiary. Many of the distinctive wrinkles on the face of today's globe, the results of continental collisions, are essentially Tertiary features. The collision of India with Asia during the Tertiary gave birth to the Himalayas, while Australia separated from Antarctica to begin its own northward drift. Wallace's line is the biological marker for the meeting point of the Australian plate with Asia.

The break-up of Pangaea was an important contributor to the post-Palaeozoic rise in biodiversity, as it created new oceanic barriers between populations of both continental and shallow marine organisms, which then underwent divergent evolution.

Hotspots and patterns of diversity

The very different faunas that occur either side of Wallace's line are the biological signature of geological history that brought together communities that were once widely separated. The reverse also happens, when new physical barriers split populations that then diverge from each other over evolutionary time. When the Zagros Mountains in Iran and Iraq were pushed skyward between 10 and 5 million years ago by the collision of the Eurasian and Arabian plates, a group of agamid lizards were separated. The colder and drier climates of the newly formed mountain tops prevented the lizard populations to the north and south from meeting. Consequently, the populations diverged genetically and gradually became separate species. Water is also a very effective barrier to migration for many species and this is why as many as one in five of the 2,300 known species of rodent is an island endemic, found only on an isolated island home. The high proportion of rodent species limited to individual islands is a consequence of evolution, aided by the inability of rodents to move between islands once chance has brought a few founders to a new place. Endemic animals and plants that have evolved on islands make a significant contribution to the total biodiversity of the planet. Oceanic island archipelagos, such as Hawaii, the Galapagos Islands and the Canaries are hotspots for endemic species.

ABOVE Turkish spiny mice are endemic to a small area of temperate woodland in southern Turkey. This limited distribution means that the wild population is critically endangered.

On a continental scale, there is a gradient in the number of species from poles to equator that is remarkably consistent across many groups of plants and animals. This latitudinal gradient in diversity was first noted by Johann Reinhold Forster, the naturalist accompanying Captain Cook's second round-the-world voyage (*see* chapter 1). Forster attributed this phenomenon to the corresponding gradient in the heating of the Earth's surface by the sun. Three regions in particular stand out on the map of plant species richness. The rainforest of Amazonia is the largest region with high diversity while the Sahara exhibits low diversity for its latitude. The flora of the Cape of South Africa is notable for its unusually high diversity compared with other regions outside the tropics.

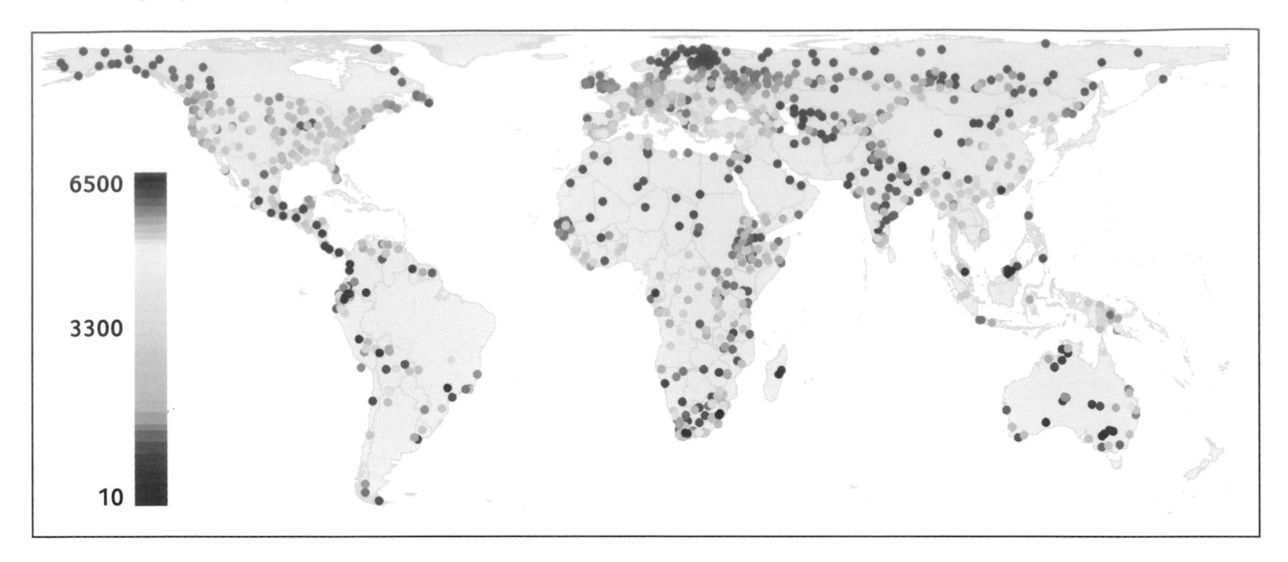

ABOVE The number of different species found in an area is known as the species richness, this is a good measure of biodiversity. The 1,032 dots on the map above indicate locations where species richness has been calculated for vascular plants. Red colours indicate high diversity, while blue colours indicate low diversity.

So why are these regions so different in terms of their modern biodiversity and what can this tell us about the origins of biodiversity hotspots? A number of lines of scientific enquiry can provide answers. Geological and fossil records can show how areas have changed through time and the genes of living species can tell us about the evolutionary history of the groups to which they belong. With these insights we can understand how and why the biological diversity of the planet is distributed in the way it is today.

How did Amazonia become so diverse?

The Amazon basin covers approximately 6 million km^2 (2.3 million miles^2) and contains highly productive and diverse ecosystems. It is estimated that around 10% of all species on the planet are contained within Amazonia, a higher concentration than is found anywhere else. A recent compilation of data suggest that there are around 400 different species of reptiles, amphibians and mammals, over 1,000 species of birds, around 3,000 species of fishes and, staggeringly, probably more than 40,000 different species of plants. Species new to science continue to be discovered within Amazonia so these numbers are always moving upwards. The dominant vegetation type is moist evergreen forest, more commonly known as 'rainforest'. Rainforest is often depicted as a large homogeneous green area on vegetation maps. However, there are huge differences in the species composition of the forest from place to place within the Amazon basin. It is more accurate to imagine this region as a patchwork or mosaic of different vegetation types. The reason for these variations is the mixture of differences found across this vast region in climate, soil, geology and topography.

In places, the evergreen forest of western Amazonia contains over 1,000 different species of plant in only a quarter of a square kilometre (a tenth of a square mile). These lush forests are home to many species of animal such as the spider monkey and boa constrictor. 'Várseas', or flooded forests, are found along the main water courses.

Within this seasonally wet habitat live capybara (the world's largest rodent), jaguar and peccaries. In permanently wet areas there are palm swamp forests. In contrast, where there is a shortage of water for a few months of the year either dry forest or savannah can develop. In dry forests many trees conserve water in the dry season by losing their leaves. Savannah grasslands, such as those in Guyana, can also support a high diversity of animal life that includes species also found in the forests such as giant armadillos, anteaters and tapirs.

Patterns of animal diversity within the Amazon basin led ecologists to suggest that high biodiversity might have been promoted by past global climate change during the Quaternary (the last 1.8 million years). Through the Quaternary the globe fluctuated between cold, glacial conditions and interglacial warm periods, like the present. These glacial–interglacial cycles resulted in the expansion and contraction of ice sheets at high latitudes and on mountains. However, the effect in lowland tropical regions remains a subject of conjecture. During glacial periods Amazonia may have become sufficiently dry to cause the fragmentation of the forest by savannah grassland. Patches of forest, isolated from similar fragments in a sea of grass, could have provided island-like conditions in which new species arose, just like on true oceanic islands.

The hypothesis that speciation in Amazonia was driven by Quaternary climate change provided a catalyst for palaeo-ecological research as scientists sought to find fossil evidence that could validate or falsify the theory. The quest to determine the nature of Amazonia during the last glacial period has been underway now for 40 years. Fossil evidence and climate–vegetation models have not supported the theory that

BELOW AND ABOVE Capybara live in wet areas of Amazonia while spider monkeys inhabit the forests.

the forest was fragmented by savannah during the last ice age. Instead data point towards a contraction of the forested area at its margins and a change in forest species composition in the core area. So, if Quaternary climate change cannot explain the high biodiversity of Amazonia, what is the explanation?

Clues to the origin of Amazonia's modern pattern of biodiversity came from the reconstruction of evolutionary trees. The evolutionary history recorded in the genes of a variety of living organisms, such as parrots and reptiles, suggested that many species originated in Amazonia during the Miocene period (between 23 and 8 million years ago). The Miocene was warmer than the present and consequently sea level was much higher than today. Indeed, examination of the geological record for the Miocene in South America reveals that sea levels were so high that much of the continent was inundated and the present land mass would have been dissected by inlets from the sea. The Miocene 'sea-ways' would have acted as significant barriers to organisms over millions of years and were undoubtedly responsible for separating populations that evolved into separate species. Comparison of the location of the 'sea-ways' and the modern genetic groups of organisms suggests that we can still see those patterns in the diversity of today.

So, a large part of the diversity within Amazonian animals and plants is the result of evolution in a series of areas isolated from one another by sea during the last 23

BELOW Higher sea levels during the warm Miocene created major geographical barriers across South America. The genetic relatedness for modern populations of reptiles, birds, and rodents follow roughly the spatial pattern created by the Miocene seaways.

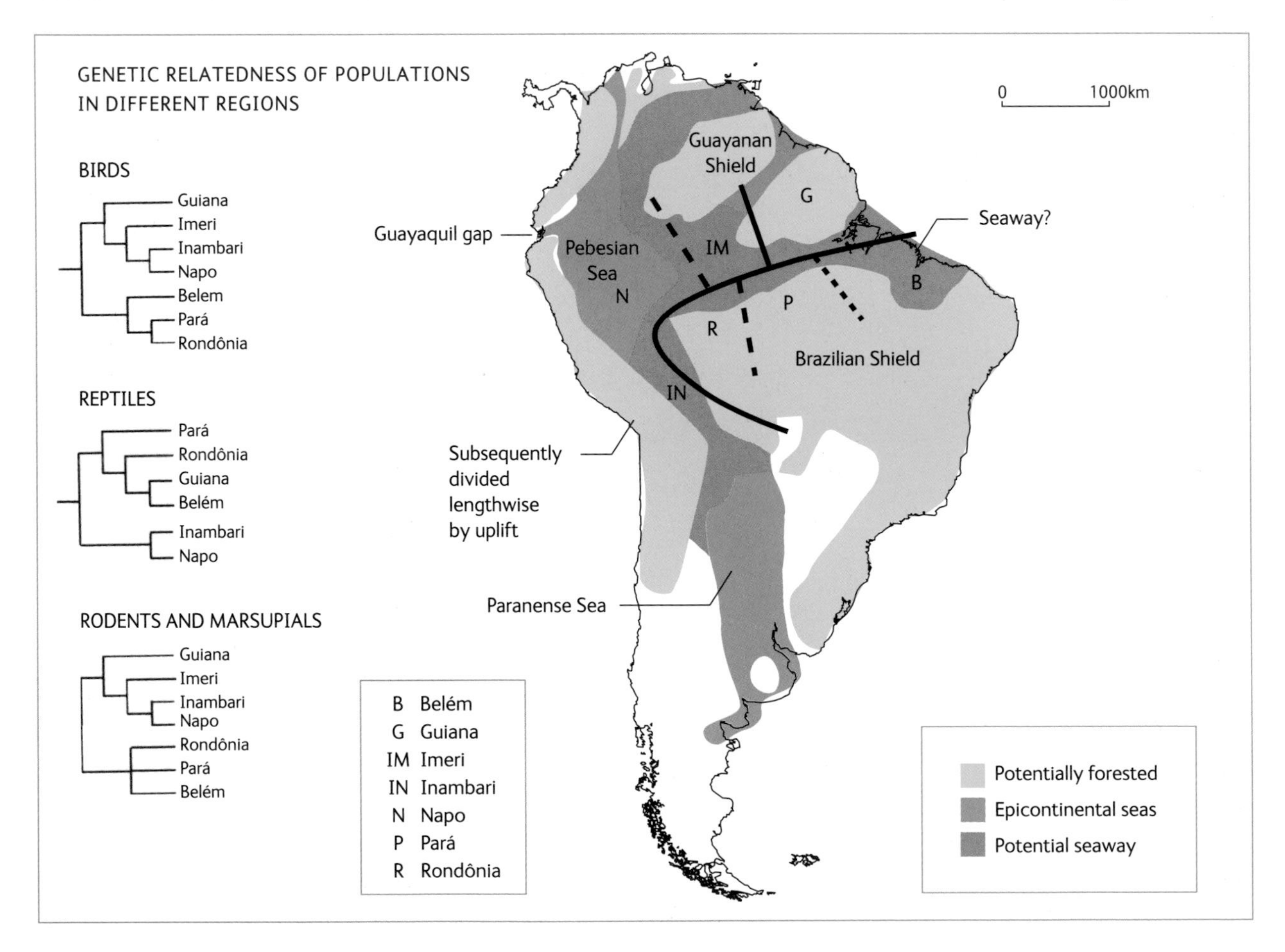

ABOVE The earliest archaeological evidence for human occupation from the Amazon Basin is from the Caverna da Pedra Pintada in Monte Alegre, Pará state (Brazil).

LEFT The cave is surrounded by evergreen rainforest today.

million years or so. The legacy of this geological and evolutionary process can still be seen today in the patterns of diversity across the Amazon basin. More recently, in the Quaternary, this highly diverse region has been joined together thanks to lower sea levels and it has become mixed into a patchwork quilt of astonishing variety. Another important event took place at the end of the last ice age: humans arrived.

The oldest dated human remains from within the Amazon basin are probably around 12,500, but may be as much as 16,000, years old. These come from the Caverna da Pedra Pintada at Monte Alegre in Brazil. By around 8,000 years ago evidence from

ABOVE Rock art from Serra da Lua provides evidence of early human activity.

archaeology and fossil charcoal indicates that human populations were widespread across the Amazon basin. These early people probably lived mainly along river courses and around lakes. Living on water courses would have had two major advantages. First, they could obtain fish as well as hunting in the forests and second, it would have been much easier to travel along the waterways compared with hiking through dense jungle. Estimates of the size of the human population within the Amazon basin before the arrival of the Spanish, in AD 1492, vary from 1 to 10 million, but given the vast size of Amazonia even if the highest estimates are correct the population densities will have been, by modern standards, comparatively low. The impact of these peoples on the environment is therefore likely to have been limited to localized clearance and hunting. Their hunting may have eliminated some large animals, but their effect on the landscape as a whole cannot have been anywhere near the scale of modern human destruction.

Why is the Sahara species-poor?

The rainforests found in Africa are highly diverse in comparison to most ecosystems on the planet but, when compared with other forests elsewhere in the tropics they are relatively species poor. For example, tropical forests in Africa contain only around half as many plant species as their counterparts in South America. In northern tropical Africa the Sahara, a desert, stands out even further as having especially low biodiversity for a tropical region. So why are both high and low biodiversity found in tropical regions?

There are two fundamental differences between the African and South American tropics. First, they cover very different latitudinal ranges. Both continents have significant landmass in the southern hemisphere (across the tropic of Capricorn, 23°S) but Africa has significant land mass as far as 30°N, whereas South America only reaches 10°N. The second major difference is climate. Tropical South America today contains many more areas that are hot (mean annual temperature more than 26°C or 78.8°F) and wet (mean annual rainfall more than 3,000 mm or 120 in per year) throughout the year. In addition, Africa has a long history of human occupation. Hominids (the group of primates to which we belong) evolved in Africa over the last *c.* 2.6 million years with groups of Palaeolithic hunter-gatherers occupying the central Sahara region as long ago as 200,000 years. In contrast, in South America the earliest physical evidence for modern humans in the Amazon basin is around 16,000 years ago.

Archaeological evidence from early humans in Africa reveals that areas which are occupied today by almost empty desert were once full of life. Writings from Egypt and cave paintings from western North Africa before 5,000 years ago record the hunting of game and the presence of animals such as crocodiles and elephants. The archaeology is backed up by evidence from the geological record. Ancient landforms, such as deltas in the Mediterranean and dry lake basins in the Sahara, show that what is now desert was much wetter between 10,000 and 5,000 years ago. Fossil pollen records from marine cores show that during the Early Holocene (*c.* 9,000 years ago) the tropical forest vegetation existed further north in Africa than it does today.

So what happened to the diverse tropical ecosystems that existed in northern Africa up to *c.* 5,000 years ago? Did humans clear the forest and destroy the environment, or did some natural event occur that altered the climate? Examination of the archaeological evidence suggests that although human populations were widespread across northern Africa, they were at low population densities. Therefore it is unlikely that humans were responsible for the loss of vegetation and drying of the whole of the Sahara. Instead, the arrow of cause and effect points in the other direction, from a drying climate to a loss of vegetation and an exodus of people as a result.

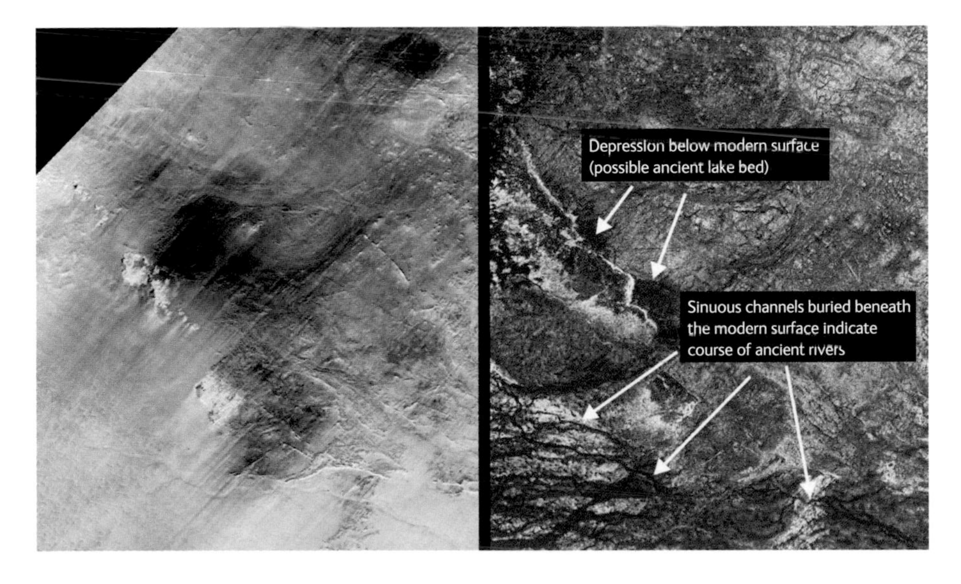

LEFT Satellite images of the Sahara show small oasis (green) and sand dune features on the surface today (left). Radar images penetrate below the surface and show evidence of ancient, wetter climates (right).

The global patterns of climate are determined first and foremost by how much heat reaches different parts of the Earth from the sun. The poles are the coldest parts of the planet on average because the tilt of the Earth's axis causes these extremities to spend half the year in total or semi-darkness. In contrast, the tropics, which lie in a belt around the equator, are strongly illuminated throughout the year. Changes in Earth's climate over time are affected by various small variations in the Earth's orbital path. These changes alter how much of the sun's energy reaches our planet and determine where that energy arrives. The superimposed climate cycles that result from orbital variation each last about 100,000 years, 40,000 years and 21,000 years. The 100,000-year cycle relates to the shape of the Earth's orbit around the sun and controls the timing of glacial and interglacial periods. The 40,000-year and 21,000-year cycles relate to the rotation of the Earth on its axis. You can illustrate this by taking a coin and spinning it on a table top. As the coin loses momentum and slows down, the axis on which it spins wobbles more and more wildly until the coin ceases to spin and falls flat. Fortunately, the friction that slows a spinning coin does not operate in space, but the Earth wobbles slightly on its axis and this changes the planet's orientation with respect to the sun. The current interglacial conditions have existed for roughly the last 11,500 years. In the tropics, fluctuations in climate are also strongly affected by variations in the 21,000-year cycle, which modulates seasonality. Seasonality is particularly important in the tropics because it controls the climate systems that deliver rain. High rainfall is required to support the highly diverse tropical forest ecosystems.

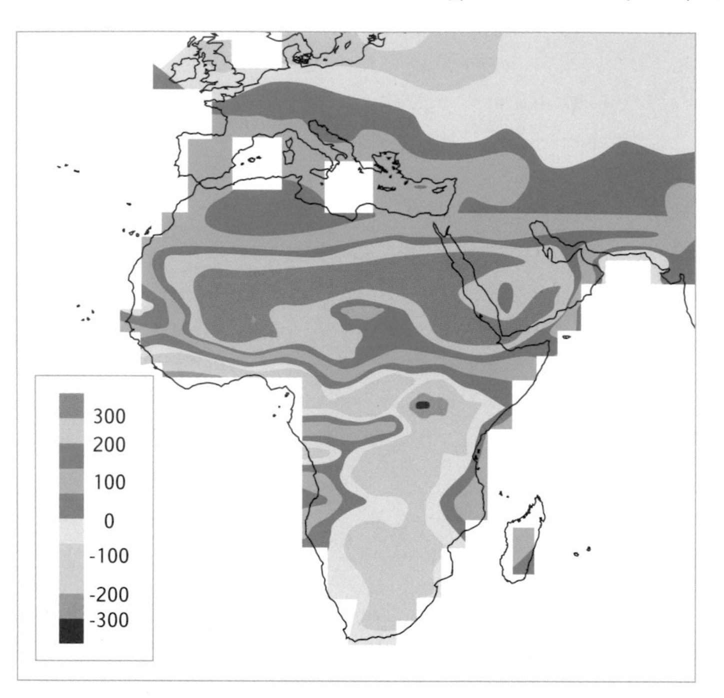

ABOVE Spatial differences in precipitation in centimetres between 8,500 years ago and present. Blue colours indicate wetter conditions in the past while reds indicate where it was drier. White areas are where there is no data.

Could planetary wobble have resulted in the transition from tropical forest to desert seen in the Sahara? Calculations of orbital conditions over the last 10,000 years indicate that there was a change 5,000 years ago, about the time that forest vegetation and people apparently abandoned the Sahara region. The 21,000-year cycle moved the region of the Earth that receives the greatest warming further south. Consequently, the climate systems that, before 5,000 years ago, generated rain over northern Africa, also moved south, reducing the amount of moisture reaching the Sahara region.

The reduction in rainfall would have been bad news for the tropical ecosystems and human populations within the region. Between 6,700 and 3,600 years ago, desert conditions were established across much of North Africa. It therefore seems likely that, not withstanding local variations, global climate change caused mainly by movements in the axis of the Earth's rotation generated the transition from diverse tropical ecosystem to desert in northern Africa. That change serves as an example of the vulnerability of ecosystems to global climate change and the detrimental effect biodiversity loss can have on human life.

Why is the flora of the Cape Floristic Region so diverse?

The plant life of the southern tip of Africa is remarkable for its extraordinary richness. The Cape Floristic Region, an area of nearly 90,000 km^2 (35,000 miles^2), is home to about 9,000 native plant species, around 6,210 of which are endemic (found nowhere else). Although occupying less than 1% of Africa's total land area, the Cape flora has around 20% of the continent's plant species. Table Mountain alone supports 2,200 plant species, more than are found in the entire UK. Much of the region is covered by a shrubby vegetation, called 'fynbos'. The Cape's diversity is doubly remarkable for being treeless and lying outside the tropics. The plant diversity of the tropics is mainly the result of the huge number of tree species there.

A number of factors seem to be responsible for the anomalously high plant diversity of the Cape, the most important of which the region does share with the tropics: neither has been scoured by glaciation. The reason the Cape escaped glaciation is that it is surrounded on three sides by ocean, which buffers its climate against extremes of temperature. The Cape did experience a drying of the climate during glacial periods, but the high mountains of the region provided moist habitats that acted as refuges for plants that could not tolerate the drier conditions, while many new species that were adapted to the dry climate evolved in the lowlands. The flora of the Cape Floristic Region therefore seems to owe its modern richness to the coincidence of a historically low extinction rate and a high rate of formation of new species.

BELOW Fynbos vegetation in the mountains of the Western Cape of South Africa, an area of extraordinarily rich plant biodiversity.

4 What is a species?

CHAPTER

MORE THAN 150 YEARS HAVE PASSED SINCE the publication in 1859 of Charles Darwin's revolutionary book *On the Origin of Species by Means of Natural Selection*, so you might wonder why we should still be asking what a species is. At first sight, the answer to the question in the title is obvious. A species is a distinct group of organisms; if organisms look different from each other, they are different species. However, there is a flaw in this argument. Think of small brown birds – you may know that there are several different species, but can you tell them apart? Where do we draw the line between mere varieties of a species and distinct species? Charles Darwin was able to answer this question because he realized that all organisms, of whatever species, have common ancestors from which they have ultimately evolved. The difference between varieties and species, he said, is only a matter of degree:

> *'Hereafter we shall be compelled to acknowledge that the only distinction between species and well-marked varieties is, that the latter are known, or believed, to be connected at the present day by intermediate gradations, whereas species were formerly thus connected.'*
>
> Charles Darwin, *On the Origin of Species* 1859 (p.485)

Darwin's phrase 'intermediate gradations' refers to the variety that is to be found in every population of living organisms. Just think of the variety of characteristics to be found among the students in any classroom or among the co-workers in any workplace. That kind of variation is also to be found in most populations of animals and plants and as a general rule the deeper you look, for example at biochemical and genetic characteristics, the more variation there is. Even populations of small brown birds vary if you look closely enough. Inherited variation is the raw material of evolution out of which new species are forged.

Evolution by natural selection and the origin of species

Charles Darwin (1809–1882) and Alfred Russel Wallace (1823–1913) are justly famous for having independently hit upon the theory of evolution by means of natural selection. In a nutshell, the theory notes that in nature the *potential* for relentless growth of populations is limited by shortage of food or suitable living-space, disease, predation and other such threats. Given the natural variation seen among organisms, those

OPPOSITE Two pinned specimens of the Spanish moon moth, *Graellsia isabellae*. Although there are visible differences they are both the same species. The upper one is male and the lower one is female.

ABOVE AND RIGHT The land iguana, *Conolophus subcristatus*, (above left) and the marine iguana, *Amblyrhynchus cristatus* (above and below right) have diverged from their common ancestor and occupy different habitats in the Galápagos islands. There is variation amongst marine iguanas as the colour differences between the two individuals shows.

possessing traits that confer advantage in this 'struggle for existence', as Darwin called it, are more likely to survive and reproduce than those not so endowed. If such traits are inherited, the advantage they confer will cause these traits to become increasingly frequent in succeeding generations.

Given enough time, this simple process can generate profound changes by tiny steps. We now also know that the reservoir of heritable variations upon which natural selection acts is kept stocked up by genetic mutations – chance alterations in genes – that give rise to new variants. Hence populations of organisms continuously adapt to the circumstances in which they live. Evolution by natural selection is still recognized today as the only process by which species become adapted. Without variation in populations of organisms there would be nothing on which natural selection could act, but it is the presence of variation that makes for difficulty in defining each species.

Ancestry

Defining a species does more than just give organisms names. Some of the characters of the species will be unique; some will be shared with other species. Defining a species therefore also defines its position in relation to others and hence starts to expose its ancestry. Darwin and Wallace realized that natural selection could not only explain adaptation, but could also help to explain the dazzling diversity of life. Different populations of the same species may be subject to contrasting conditions and so

evolve divergently, eventually becoming distinct species in their own right. Continued branching and divergence of lineages through long spans of time combined with the extinction of intervening branches lead in turn to the emergence of distinct clusters of related species. The latter correspond to the natural groupings that have long been recognized in the classification of organisms, such as mammals, birds, snails or flowering plants, which are all ultimately related in a single grand genealogy, or 'tree of life'. Darwin graphically illustrated this concept in the one figure that he included in his great book *On the Origin of Species* (1859).

ABOVE Branching descent, as illustrated by Charles Darwin in *On the Origin of Species*. The horizontal lines labelled with Roman numerals represent intervals of thousands to millions of generations, over which time new species evolve and many species become extinct.

HOW MANY SPECIES ARE THERE?

It is very difficult to put a figure on the number of species on the planet. We know that over 1.4 million have been described and we also know that there are many undescribed. The best estimate is that there are at least 10 million and probably more, so maybe as few as 10% of species have been described so far. A recent re-assessment of the amphibians of Madagascar, published in 2009, suggests on the basis of genetic evidence that the 244 species described so far represent only a proportion of the total number present on the island. The true total lies between 373 and 465. As only a proportion are described and named, species may be being lost before they have been recognized.

CONSERVATION OF 'LIVING FOSSILS'

A few living species are remarkable for being the lone survivors of long-extinct lineages and, it can be argued, deserve preservation on these grounds alone. For example, the Wollemia pine, *Wollemia nobilis*, has fossils that are 200 million years old and was thought to be long-extinct until the discovery of a small population in a canyon near Sydney, Australia in 1994. The coelacanth, *Latimeria chalumnae*, is a lobe-finned fish belonging to the lineage of fishes from which four-legged land animals evolved. It was thought to have gone extinct 80 million years ago, until it was found off the east coast of South Africa in 1938. The duck-billed platypus, *Ornithorhynchus anatinus*, is another example; the earliest fossil discovered is 112 million years old.

BELOW A duck-billed platypus.

Pinning on a name

Darwin said that species are separated by discontinuities or gaps in variation. How do such gaps arise? If all life evolved from common ancestors, why isn't all life just one big species? One answer to this question is that populations become isolated from one another, for example on different islands or in remote patches of habitat, and over many generations they diverge to a point where members of the different populations are unable to interbreed if they are brought together again. This reflects the oft-used 'biological' definition of a species, which states that:

> *Organisms belong to the same species if two members can interbreed and produce viable offspring that can themselves breed.*

This definition generally works well but it does throw up a problem. Our world is full of millions of different organisms – plants, bacteria, animals – and it would be quite

ABOVE Males and females of four species of butterfly, collected by Alfred Russel Wallace. They show how different sexes of the same species might well be thought to be from different species.

impractical to conduct breeding experiments to check if those that look different are actually separate species. The two moths pictured on p.42 are similar, but you can pick out some small differences, for example in the antennae. On their appearance you would probably say that they are the same species. The 'biological' species definition cannot be applied because they are pinned specimens from 1981. In fact, they are a male and a female of the same species of moon moth. But what about the butterfly specimens collected by Wallace himself? The appearance of all of these is very different, but they are actually male and female pairs of four species.

So the first, and oldest, way of identifying different species, which has been to look for similarities in external appearance, has to cope with sexual differences – called sexual dimorphism – as well as other variations in appearance, such as colour. The importance of being able to identify which species a particular organism belongs to is that natural selection operates within populations and new species evolve from existing species.

WHAT'S IN A NAME?

A species needs a name to give it recognition, a description to allow others to see why it is considered a species and a type specimen for reference. These requirements were established long ago in the eighteenth century when Latin and Greek were the languages of learning, so species names were written in these languages then and (with a few exceptions) remain so today. Species that are similar are grouped together into a genus. So each species has the name for the genus, e.g. *Lacerta*, plus its own species name, e.g. *viridis*, for the common green lizard but *Lacerta agilis* for the closely related but distinct sand lizard. Quite often, the species name is partly descriptive, for example the green lizard is *Lacerta viridis*, *viridis* being the Latin for green.

The cobras are a group of snakes with a characteristic appearance. There are 22 species in the genus *Naja*. The picture shows *Naja ashei*. The species was first described in 2007 by two scientists, Wüster and Broadley, who named it after James Ashe, the first scientist to identify these snakes as a distinct species. It is not considered proper to name a species after oneself, so had Ashe himself described it he could not have given it the species name *ashei*. Interestingly, the snake that we call the king cobra is not in the same genus. This highly venomous, very large snake is called *Ophiophagus hannah*. Who was Hannah and why did she have a cobra named after her?

With millions of different species to name it is inevitable that some strange ones emerge, for example the snail from the Fijian island of Mba that is called *Ba humbugi* or the Brazilian pterosaur fossil called *Arthurdactylus conandoylensis* – a reference to the author of the novel *The Lost World*.

BEOW The cobra, *Naja ashei*, one of twenty-two species in the genus *Naja*.

There are other definitions of a species. A recent twenty-first century list contained six different categories of definition, so the question 'what is a species?' turns out to be far from easy to answer. Genes, morphology, ecology and geography all influence species definitions. In practice, however, biologists work with species all the time and the scientific names that are used to describe distinct species generally do just that. Look at two members of the crow family. The all-black carrion crow is quite distinct in appearance from the hooded crow, which has large areas of grey. So it is easy to distinguish them in the field using visible characters and providing these characters are stable, there is no confusion.

These two species occur in Europe and eastwards into Asia. In Europe, each species occupies a different but contiguous geographical area, as the map shows. Along the boundary line, hybrids are found but they do not spread very far. At its widest point that zone is 160 km (99 miles) across. The two species remain distinct over almost all their range. Interestingly, as the climate has changed, the carrion crow has extended its range. The hybrid zone has moved northwest in Scotland, but not increased in size.

The breeding success of pure-bred crows and of hybrids between the two species is difficult to assess. A study of breeding in crows of different ancestry found that hybrid females were less successful in rearing chicks to fledglings and that carrion crow females were less successful when in the hybrid zone, although the numbers of observations were

ABOVE The hooded crow, *Corvus cornix* (above) and the carrion crow, *Corvus corone* (below) are quite distinct in appearance.

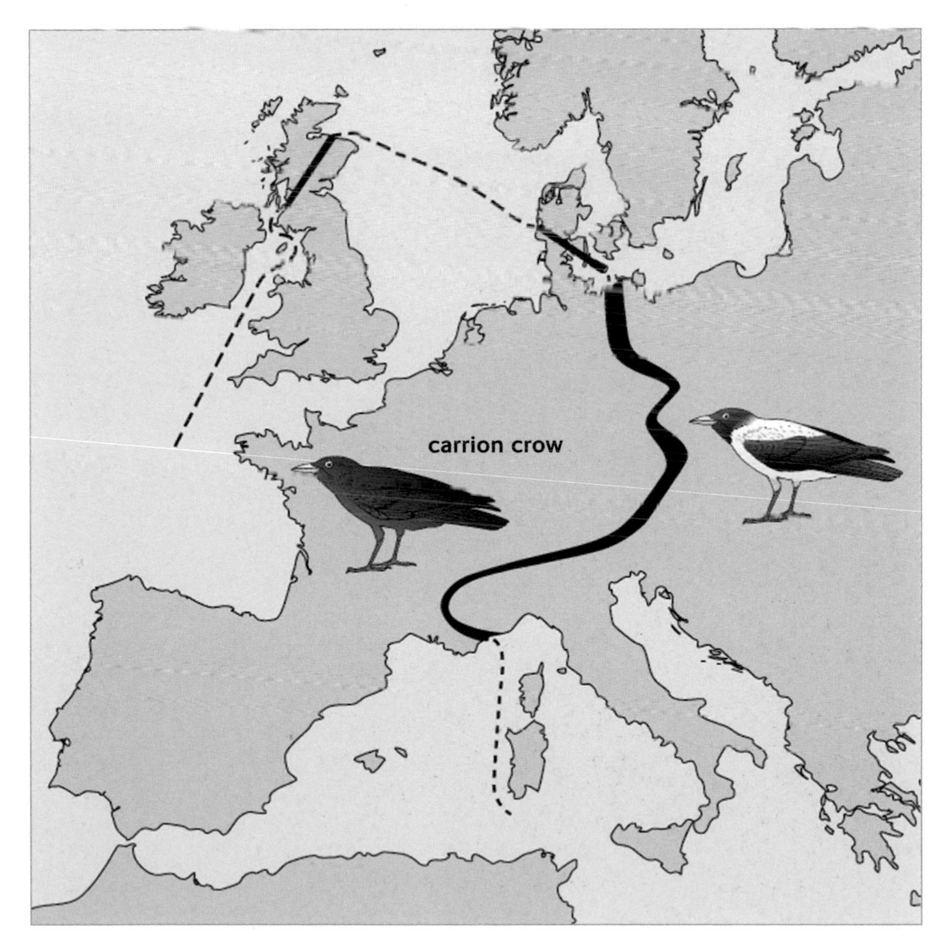

LEFT The distribution of the hooded crow and carrion crow in Europe. The black line shows the hybrid zone between the two species.

very small. Outside the hybrid zone, carrion crows were significantly more successful than hooded crows, producing a mean number of 3.1 fledglings per nest compared with a mean of 2.6 for female hooded crows. Inside the hybrid zone, carrion crow females were less successful than outside the zone with a mean numbers of fledglings of 2.5.

Carrion crows and hooded crows are distinct species and, although they do interbreed where they overlap, the hybrids are not successful enough to expand their range and supplant either of the pure-bred parental species. It looks as though reduced breeding success within the hybrid zone is part of the mechanism that maintains the boundary between the species. At a practical level, therefore, the species distinction works and we can use the plumage colours to differentiate between the two kinds of crow. Such distinctions are easy where there are clear visual and geographic differences.

Now consider some amphipods, small shrimp-like animals commonly known as sand-hoppers or sand fleas, some of which live in sand both on beaches and offshore. They are very similar and it is necessary to find distinguishing features that are constant and recognizable. Small structural characters have to be examined. One of the characters that allows us to distinguish *Bathyporeia elegans* from *Bathyporeia pilosa* for example is the presence of a tiny tooth on one of the skeletal plates and the pattern of spines at the margin. This is easier to see in drawings. These characters may have no obvious role in the

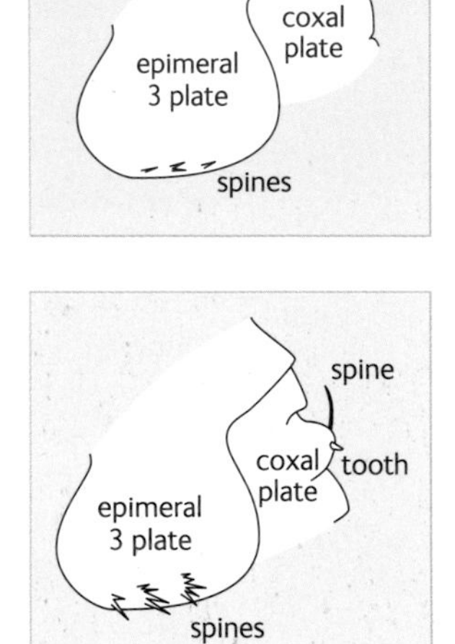

ABOVE Drawing of two skeletal plates of *B. pilosa* (above) and *B. elegans* (below) showing the small tooth that is one of the characteristics used to distinguish between the species.

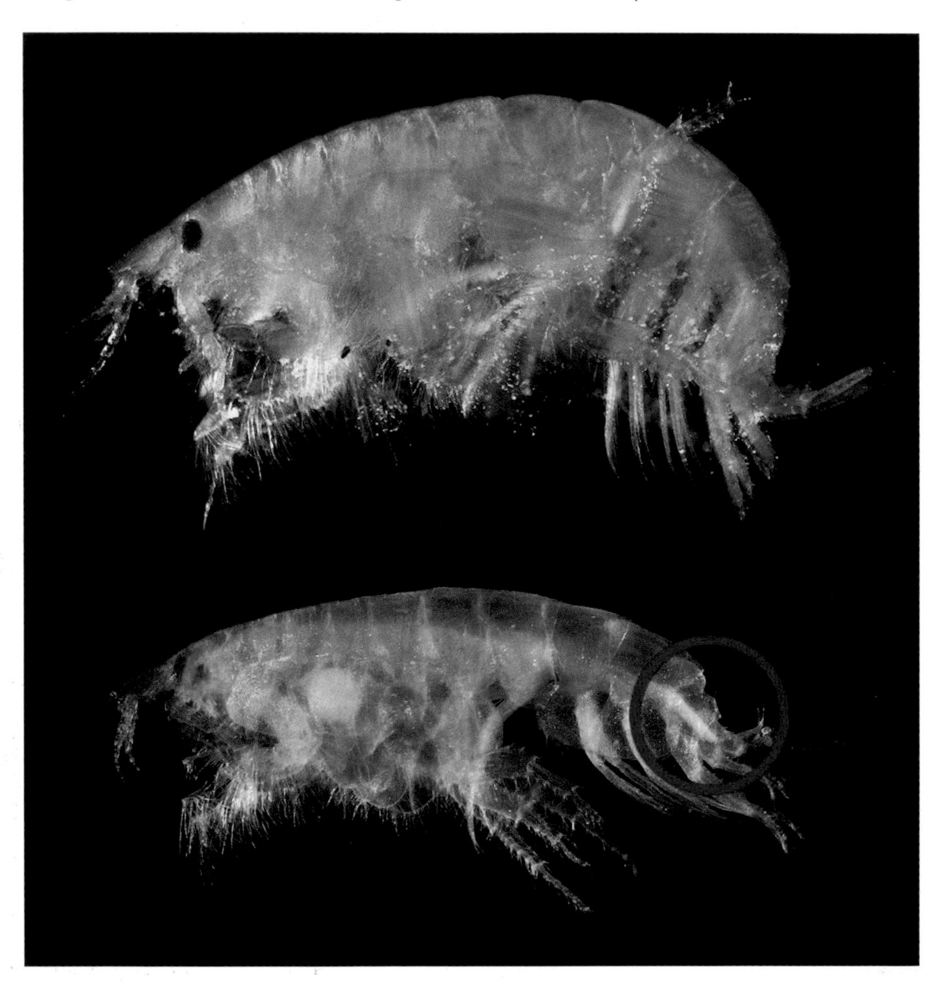

RIGHT *B. pilosa* (above) and *B. elegans* (below). The red circle shows the area illustrated in the drawings above.

life of the animal, but provide some of a number of small differences that cumulatively enable us to distinguish between the two species. The distribution of these two species is slightly different, one being found mostly offshore while the other is also found in intertidal sand, possibly because they flourish in water of slightly different salinities.

From the examples above, you can see that while species are *defined* by whether they can interbreed and produce viable interbreeding offspring, they are *recognized* in a quite different way by being possessors of a particular suite of characters. Often these differences are visible, but they could be audible, as in the case of two species of bush cricket that look identical but have very different songs.

Some species are cryptic

Some species of bumblebee are very difficult to identify, even for an experienced entomologist. A group of bumblebee species belonging to the genus *Bombus* are widespread and commercially important as pollinators, but the number of species in the group is still not clear and so difficult is it to identify some species that *Bombus lucorum* has been described under 186 different names. However, analysis of genetic differences has identified distinct groups within *B. lucorum*. It is possible to distinguish four species that are very difficult to separate on appearance but absolutely clear-cut when examining the genes. These are cryptic species.

As genetics becomes more widely used to identify species, more cryptic species will come to light. For example, shearwaters, a widespread sea bird, are almost certainly a complex of cryptic species. Other birds, some amphibians, insects and worms are suspected of having complexes formed of cryptic species. Identifying cryptic species has important implications for conservation.

BELOW *Bombus lucorum*, the white-tailed bumblebee, visiting a flower of Himalayan balsam, *Impatiens glandulifera*.

ABOVE Giraffes on the banks of the Chobe River in Botswana. Depending on the region, the patterns on the giraffes skin can differ quite dramatically.

Giraffes are large browsing mammals that roam widely in Africa. Living in large groups with a range of up to 300 km (186 miles) you would predict that populations were unlikely to be geographically or behaviourally isolated with plenty of interbreeding within the species. Giraffes are generally considered to belong to one species, *Giraffa camelopardalis*. There are quite large variations in the patterns on the skin and some patterns are linked to geographical regions. Where the pattern was distinctive, the populations were assigned to a sub-species, for example *Giraffa camelopardalis rothschildi*. A sub-species is defined as a geographically distinct population that has recognizable differences from other members of the species. Sub-species do interbreed but are normally isolated. Over time, subspecies may evolve into separate species because so many differences accumulate that hybrids become less viable or less fertile than pure-breds, as in the case of the crows. A recent study of giraffes using mitochondrial genes has completely changed our view of the species by demonstrating the existence of at least six separate species of giraffe, some with distinct genetic sub-groups, giving a total of 11 genetically distinct populations. This discovery has important implications for those studying evolution and natural selection. It also has profound implications for conservation because efforts now must be much more targeted on separated, genetically distinct populations, such as the 100 individual giraffes at a single site in Niger. The Red List that identifies species threatened with extinction places giraffes as being at lower risk on the basis that there are an estimated 110,000 individuals, all of one species. The picture is different now. Clearly some of the genetically distinct groups should be on the critically endangered list and managed as separate species.

New species from old

Even while existing species are being described and catalogued, new ones can appear – and appear rapidly. An estuarine cord grass from the US East coast, *Spartina alternifolia* was introduced into San Francisco Bay where there is a native species *Spartina foliosa*. Not only did the introduced species spread rapidly but it also hybridized with the native species and the hybrids spread widely too. In the nineteenth century the same species, *S. alternifolia*, was introduced into Europe, where there was only one native species, *Spartina maritima*. The two species hybridized, producing a sterile, perennial hybrid. Like many plants, cord grass can propagate by asexual reproduction; clumps break apart and each establishes a new plant (gardeners use similar principles to propagate artificially produced hybrid cultivars).

However, fertile hybrids subsequently appeared that formed a new species, *Spartina anglica*, which now does not interbreed with either parental species. The new species does better in estuarine conditions than either of its distant parents and it has spread into northern Europe, Australia, New Zealand and North America, earning it a place on the list of 100 of the 'world's worst invaders'. All of this evolution has occurred in just 140 years!

BELOW *Spartina anglica*, the common cordgrass, growing in an estuarine habitat.

Conserving species or conserving genes?

BELOW The Lord Howe Island stick insect, *Dryococelus australis* (above) and the New Guinea spiny stick insect, *Eurycantha calcarata* (below). Although only distantly related, the two insects look similar and both have the unusual backwardly pointing spines on the hind legs.

There is continuing debate about where scarce resources should be concentrated to conserve threatened species. Should we protect, if we can, any species threatened with extinction or should we consider the position that species occupies in an evolutionary tree? The choice is between conserving species richness and conserving evolutionary diversity. Consider the Lord Howe Island stick insect, which was thought to be extinct following the disappearance of the population on the remote coral island 600 km (373 miles) east of Australia. This large flightless insect has a backwardly curved claw on the hind leg and because it closely resembles other stick insects found in Southeast Asia, was presumed to be closely related. Here then is an argument that extinction, although regrettable from the species richness point of view, did not represent a loss of genetic diversity because close relatives were flourishing elsewhere.

Recently, on a tiny island called Ball's Pyramid, about 20 km (12 miles) from Lord Howe Island, a population of maybe 30 individuals has been found. Captive breeding is being undertaken. Analysis of mitochondrial genes shows that this stick insect actually has a quite separate ancestry from the groups it resembles – it just looks like them and is a great example of convergent evolution. The studies suggest that it separated from the most recent common ancestor over 22 million years ago. As Lord Howe Island and Ball's Pyramid were formed 6.4–6.9 million years ago, its ancestors must have lived elsewhere for a period and, presumably, those populations have now vanished. So the isolated geographical and evolutionary position of the insect, coupled with the long period

separating it from any other known species, make the conservation of this species much more important from the evolutionary than the species richness point of view.

How the insect colonized the islands is unknown. If captive breeding fails, perhaps because there is insufficient genetic variation in the three insects that founded the laboratory population, would anybody risk taking more from the wild? The Lord Howe Island stick insect may be a small-scale conservation issue, but it highlights the difficulties faced in conserving any population as numbers of individuals fall. It also shows the value of genetic studies in establishing how distinct a species is.

The vexed problem of the red wolf, *Canis rufus*, illustrates the interaction between species definition by visible characters and molecular data. The red wolf appears to be a distinct species and is desperately in need of conservation. It is officially classified an endangered species, as its numbers fell to just 17 individuals in the 1970s and it became extinct in the wild in the mid-1980s. A captive breeding programme has been established for 30 years and there are now over 100 red wolves living in the wild in the USA and over 200 in captive breeding programmes. Yet genetic data suggest that it is actually a hybrid between two other wolf species and so, some would argue, does not need expensive conservation measures. Analysis of mitochondrial genes from extant red wolves found that they were identical to those from coyotes. Studying samples from historical specimens preserved in museums revealed only two types of mitochondrial genes: grey wolf and coyote, with no distinct red wolf mitochondrial genes. Is it possible that there was a hybrid zone between grey wolves and coyotes and that the red wolf is a hybrid between these two species? If so, perhaps it would not require conservation and captive breeding programmes.

It is fair to say that there is dispute over the genetic evidence and that the status of the red wolf is the subject of a complex discussion, but the debate illustrates the very real decisions that have to be taken about how scarce resources are used in conservation and how it matters what you designate as a species. The battle lines are drawn. Does the origin and taxonomic status of the red wolf conflict with its conservation plan? More generally, genetic diversity is a component of biodiversity, but how important is it that this is preserved? Genetics has the answer.

5 CHAPTER Genes, genes, genes

THE NATURAL HISTORY MUSEUM IN LONDON is a treasure-trove of biodiversity, pickled, pinned, pressed and preserved by collectors whose discoveries have been finding their way into the museum's cabinets for four centuries. Many of those cabinets are older than the museum itself, but a relatively recent acquisition is a set of 28 of Alfred Russel Wallace's glass topped-wooden drawers, which contain the great naturalist's pinned insect specimens. They formed a small personal collection that his descendants sold to the museum as recently as 2002. All of the drawers are remarkable, containing the gems of a lifetime's collecting in the tropics, undertaken a century before its forests were destroyed by late twentieth century deforestation. Most of them illustrate the wonderful diversity of tropical insect species, but a couple display variation within species, in one case between the two sexes of moon moth and in another the individual variation to be found in dead-leaf butterflies.

BELOW Four dead-leaf butterflies illustrating individual variation within a species.

The butterflies are camouflaged and look like inedible, dead leaves in a manner that Wallace described as '...the most wonderful and undoubted case of protective resemblance in a butterfly...'. When resting '...the little tails of the hind wing touch the branch, and form a perfect stalk to the leaf...the irregular outline of the wings gives exactly the perspective effect of a shrivelled leaf.'

Although every dead-leaf butterfly was perfectly camouflaged, each was as subtly different as dead leaves are from each other, so that '...out of fifty specimens no two can be found exactly alike, but every one of them will be of some shade of ash or brown or ochre, such as are found among dead, dry, or decaying leaves.' There were even individuals with simulated holes in their wings or with markings giving the appearance of the fungi that colonize the leaves of tropical trees. As the dead-leaf butterflies demonstrate, biodiversity is not just about numbers of species, it is also about how much variation there is *within* species. All species vary to some degree, and a significant proportion of that variation is usually inherited. Inherited variation is in the genes.

What are genes?

Genes are genetic instructions written in a sequence of molecular letters that are linked together in very long chains of DNA that, with some structural and control molecules, form chromosomes. There are just four letters, representing nucleotides, in DNA's alphabet and they are A, C, G and T. If we think metaphorically of a gene as being a one-sentence instruction, then a chromosome is like a chapter composed of many

OPPOSITE The local population of the Florida panther was saved from the ill effects of inbreeding by the introduction of animals from Texas (see p.61).

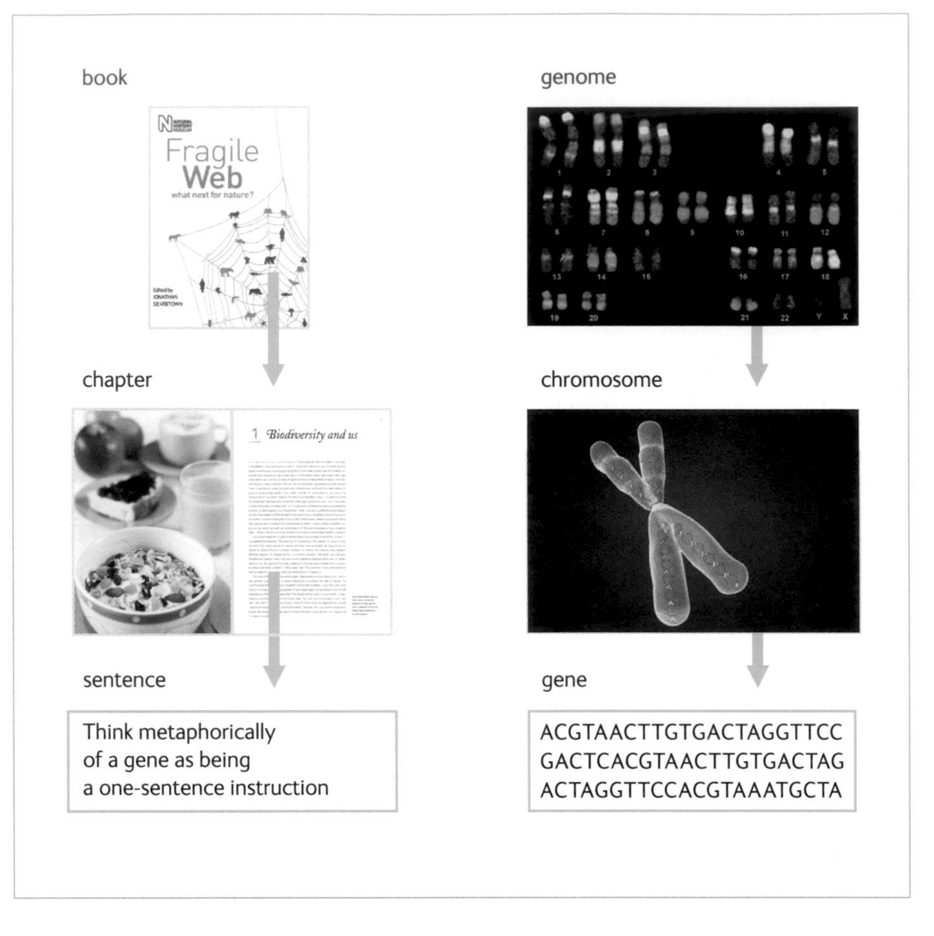

RIGHT The structure of a genome is like a book composed of many chapters (chromosomes), containing sentences (genes) written with just four letters (the letters of the DNA code).

sentences. A whole genome is then a book containing many chromosome 'chapters'. In the book of the human genome, there are 23 chromosome chapters. Most cells of your body contain two copies of each chromosome, one copy derived from your mother, the other from your father. If you yourself have children, only one copy of each of your 23 chromosomes is transmitted to them. (If children received both copies from each parent, chromosome numbers would double every generation, from your 23 pairs, to their 46 pairs, to their children's 92 pairs and so on.) The reduction from 23 pairs of chromosomes in your body cells to the single set of 23 that you pass on to your children takes place during the formation of gametes (the collective term for sperm and eggs).

The curious thing about inheritance is that diversity and resemblance both spring from the same source: the genes. Offspring resemble their parents but they are not identical to them. Diversity among offspring comes from the fact that many genes occur in slightly different versions, called alleles. Alleles arise when mutation changes the letters of the coded messages in the genes. Sometimes such alterations fatally damage the gene's function, and sometimes there is no visible effect; rarely, but most importantly for evolution, mutations may be beneficial for the individual. Although new mutations are relatively rare, the alleles they give rise to are numerous because non-harmful mutations have accumulated over the very long time of evolution. Many, perhaps most, mutant genes are much older than the modern species that carry them and were present in those species' ancestors.

Each of us can have a maximum of only two different alleles for any one gene (one allele on each of the chromosomes in a pair), but in a population as a whole there can be many more than two. For example, the blood groups that determine compatibility between donor and recipient in blood transfusions are determined by three alleles A, B and O which all belong to a single gene. Sequencing of the ABO gene using DNA taken from 55 different people discovered more than 80 different mutations hidden within the three alleles, though many of the mutations were neutral, having no obvious effect. Neutral mutations are so abundant in many animal and plant populations that they can be used to create a genetic fingerprint that uniquely identifies an individual from a sample of his or her DNA. DNA fingerprinting is widely used for this purpose in forensic science, including tracing illegally traded wildlife. When all the genes and all their different alleles are taken into account, we are each a unique combination of our parents' genes; hence each generation delivers family resemblances with a twist of variety.

What do genes do?

Genes are the carriers of inherited variation and most importantly, only variation that is inherited can provide the raw material for evolution. Genetic diversity is a measure of the number of alleles (alternative DNA sequences) that there are within a population of a species. The existence of cryptic species, like those found in bumblebees or giraffes (*see* chapter 4), illustrates the fact that a great deal of genetic diversity is hidden from normal view, but it can be revealed by analyzing and comparing DNA sequences.

The number of genes in an organism ranges from about 1,000 in small bacterium to more than 700,000 in many flowering plants. The amount of information contained in these genes is phenomenal – as a human, for example, you have about 3 billion nucleotides in your genome, which are copied into nearly all of your estimated 100 trillion cells! Although we do not know the function of all genetic material (or even

LEFT Heart muscle, seen here greatly magnified, is made largely of proteins, and the instructions for their manufacture are in the genes. Proteins make up much of an organism's biochemical machinery and structure such as skin, hair, blood cells, the brain and components of the immune system.

if it all has a function), we do know the roles of many genes. Ultimately, genes make proteins, and proteins in turn make up much of an organism's biochemical machinery and structure such as skin, hair, blood cells, heart muscle and components of the immune system.

The immune system provides an important example of genetic diversity that is not reflected in physical appearance. Consider, for example, the winter 'flu season. During this time, not everyone who comes into contact with the virus becomes ill. The outcome depends partly on their genes – some combinations of genes provide better resistance to a particular virus (although not necessarily all viruses) than others. However, resistance to pathogens is only one reason why genetic diversity is important. There are several important reasons to be concerned about how much genetic variation is maintained within species.

Why does genetic diversity matter?

Most populations harbour a fairly high level of genetic diversity that has been acquired over a long period of time. The challenges that we are currently facing stem largely from the fact that in recent decades the genetic diversity of many populations has been lost because of human activities such as overhunting, or developments such as cities and farms that have fragmented wild populations of animals and plants. Does it matter that genetic diversity is being depleted? The simple answer is yes, the loss of genetic diversity is a very serious problem. In the short term, the problem is caused by inbreeding, which in the longer term puts in peril the potential of a species to evolve and adapt to a changing environment.

Inbreeding occurs when two relatives mate and produce offspring. The more closely related the parents are to each other, the more inbred their offspring. Inbreeding leads to increased levels of homozygosity, which occurs when the offspring inherit the same allele from each parent. Everyone is homozygous for many genes and this is not a problem where the alleles in question are not faulty, but problems arise if an individual has two copies of a faulty allele. Single copies of a deleterious allele (one that causes disease, death, abnormal development, etc.) may be transmitted from one generation to the next without an adverse effect on carriers because they are usually paired up with a normal allele that compensates for the faulty one. Individuals who inherit two copies of the faulty allele may suffer abnormalities, and this risk is greatly increased by inbreeding. Inbreeding that leads to a reduction in survival and reproduction is known as inbreeding depression.

Most animal and plant species have evolved mechanisms that help them avoid inbreeding. Studies have shown that a wide range of species – including birds, fish and mammals – use olfactory cues to avoid mating with genetically similar partners. The elaborate, colourful and enticing flower structures found in plants attract insect pollinators that transport pollen between unrelated individuals. As a consequence, even though most plants are hermaphrodites and have both male and female sex organs, cross-pollination enables them to avoid inbreeding.

ABOVE An isolated population of the common adder, *Vipera berus*, in Sweden, which was suffering from inbreeding depression, had its fate reversed by the introduction of several male adders from a genetically diverse population.

When inbreeding cannot be avoided, natural populations with low levels of genetic diversity have experienced inbreeding depression, which leads to abnormalities, and in extreme cases to extinction, for example in the highly inbred Glanville fritillary butterfly populations in Sweden. One of the main goals in conservation genetics is therefore to eliminate or minimize inbreeding within populations. When in the 1990s the Florida panther showed signs of inbreeding (these signs included poor-quality semen and undescended testicles), eight females from a closely related subspecies in Texas were moved to Florida. This introduction of new genes into the population had the desired effect of reducing inbreeding depression and creating healthier panthers, although the long-term survival of this species may still be threatened by the lack of suitable habitat in this highly developed State. A similar strategy of genetic rescue saved a population of Swedish adders, in which inbreeding had led to a high proportion of deformed or stillborn offspring. Twenty adult males were imported from a much larger and more genetically diverse population in another part of Sweden. After four breeding seasons, the newly introduced genes had led to a remarkable increase in the overall health of the population, which by this time included 39 healthy adult males – the highest number of breeding males in more than two decades.

Avoiding inbreeding can be a particular challenge to programmes that aim to conserve species through captive breeding. This is usually a last-resort option that aims to preserve species that are either extremely rare or have already gone extinct in the wild. Because these programmes are seldom implemented before a species has reached very low numbers, and hence has very few potential mates, there is a high potential that captive-bred species will experience high levels of inbreeding. Fortunately, there are also steps that can be taken to minimize this risk, for example scientists can use genetic data to determine how closely related individuals are to one another; managers can then ensure that matings between parents and offspring, or between siblings, do not occur. This approach has been used in a range of captively bred species including St Vincent's parrot, the Jamaican boa and the Persian wild ass. Another way to minimize inbreeding in captively bred species is through the co-operation of zoos, aquaria, game parks and other institutions that sometimes 'share' breeding adults. For example, the captive breeding programme of the golden lion tamarin, a small monkey native to Brazil, is managed as a Species Survival Programme of the Association of Zoos and Aquariums. As a result, 150 different zoos are involved in a 'group' captive breeding strategy that greatly increases the pool of potential mates for any given zoo.

Evolutionary potential

The second reason why genetic diversity is a good thing is that it increases evolutionary potential, in other words it increases the likelihood that at least some individuals within a population can adapt to changing environmental conditions. If species are to have a good chance of prospering in new environments altered by climate change, increased urbanization, new pathogens, etc., they need to have lots of different alleles, because only some are likely to be well adapted to each new environment. Over the longer term, all environments may be considered unpredictable, and therefore genetic diversity may always be necessary for ongoing adaptation and survival.

BELOW A red squirrel, *Tamiasciurus hudsonicus*, population in the Yukon, Canada, has altered the timing of its breeding season in the past decade in response to increasing temperatures.

Evolutionary potential is of growing concern to conservation biologists because of climate change. Species that find themselves in a substantially altered environment can move to a more suitable location, or they can adapt to the altered environment. Movement may not be possible in fragmented habitat, or for particularly slow-moving species. Adaptation is possible only if the species has the necessary underlying genetic variation. Some species, such as the red squirrel and the fruit fly, have already adapted to some aspects of climate change, but there is great concern that species that consist of small, fragmented populations cannot sustain the necessary genetic variation.

The importance of evolutionary potential can be further illustrated by examples from agriculture. For centuries, humans have selected crops and livestock for desirable traits such as larger and fatter pigs, and wheat that produces large, synchronously ripening seeds. This artificial (human-directed) selection eliminates 'undesirable' genetic lineages and substantially reduces genetic diversity. The wild grass teosinte (*Zea mays* ssp. *parviglumis*), for example, was domesticated into modern maize (*Zea mays* ssp. *mays*) in a process that started thousands of years ago, and which has resulted in today's genetically impoverished crop.

ABOVE LEFT Wheat and other crops are the products of centuries of artificial selection by farmers.

ABOVE RIGHT This exotic-looking corn is being used in breeding programmes to broaden the genetic base of the crop grown in the USA, which was impoverished by the widespread use of a limited number of varieties.

While in the short term there are obvious benefits to growing corn with large stalks, or breeding cows with high milk production, one risk to raising plants or animals with very low genetic diversity is disease outbreaks. Perhaps the most famous example of a crop disease is the Irish potato famine, which occurred in the nineteenth century when the fungus *Phytophthora infestans*, commonly known as blight, was inadvertently introduced into Ireland. Assisted by unusually warm, damp weather that favours its proliferation, the blight destroyed up to 50% of potato crops. Because potatoes were a staple food at that time, the crop disease caused famine that led to an estimated 1 million deaths in Ireland and 1 million emigrants from that country. More recently, in 1970 the USA lost 15% of its corn crop, worth about $1 billion, when a fungus spread rapidly across the Midwest. In both cases, very common and widespread host varieties were susceptible to the fungus in question, and the lack of genetic diversity resulted in large-scale obliteration. The potential dangers of low genetic diversity are clear and give cause for concern about the future of wild populations that have been broken into fragments.

Breaking up is hard to do

Limited genetic diversity in some modern populations can be traced back to the effects of migration since the last ice age. In the northern hemisphere, ice sheets and tundra extended much further south 10,000 years ago than they do today. As the ice sheets melted, plants and animals began to re-colonize the newly habitable land, but only a subset of individuals made the trek northwards. As a result, only a subset of the population, which carried only a fraction of the genetic diversity in the species, became the ancestors of the new colonies. To this day, more northerly populations of many species in the northern hemisphere have relatively low genetic diversity; these include

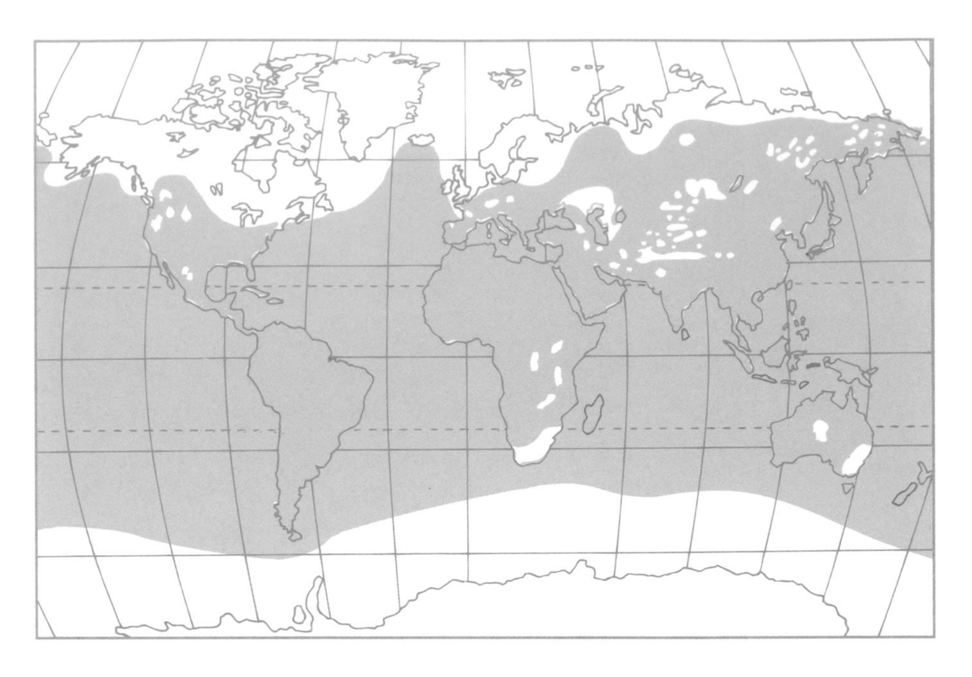

RIGHT Map showing extent of ice sheets and tundra during the last glacial maximum. The white areas on the map represent regions that were covered in either ice sheets or tundra.

North American voles, *Microtus longicaudus*, dragonflies, *Anax junius*, black spruce, *Picea mariana*, and European rock ferns, *Asplenium* spp.

Although past fragmentation of populations has had a lasting effect on genetic diversity, more recent movements of genes are generally more important because they often result in the movement of genes from one population to another. This can result from processes as diverse as pollen being transported by bees to a new destination, spiders being blown long distances on air currents, or mice travelling across land from one site to another. If a relatively small population frequently absorbs immigrants from other populations that are genetically diverse, then it may be able to maintain high levels of genetic diversity. A well-studied example is Morelet's crocodile in Belize, Central America. Morelet's crocodile is an endangered species that suffered greatly from overhunting in the twentieth century, but frequent migration of crocodiles between populations has helped to keep genetic diversity reasonably high.

BELOW Morelet's crocodile, *Crocodylus moreletii*, suffered greatly from overhunting in the mid-twentieth century, but fortunately one or two populations remained large and genetically diverse, and they have provided immigrants to other populations that have greatly reduced the potential effects of inbreeding.

Today, conservation biologists around the world are extremely worried about the effects that anthropogenic activities are having on the genetic diversity of natural populations. Habitat loss and fragmentation are the main reasons why so many species and populations are now threatened, endangered, or have already become extinct. In most cases, habitat loss is a direct result of human development. Imagine a formerly contiguous block of forest, or prairie, or sand dune, which has been repeatedly

LEFT Fragmentation and loss of tropical forest habitat between 1992 (below) and 2006 (above) caused by clearance for agriculture in the Matto Grosso, Brazil.

fragmented by roads, houses, shopping precincts, office blocks, deforestation, and so on. Many of these human developments are uninhabitable to a wide range of species that include mammals, plants, amphibians and insects. Unless individuals are capable of bypassing such developments, previously large populations now become a series of small, unconnected populations. Once movement between populations is reduced or eliminated, genetic diversity starts to decrease. Populations are then more likely to go extinct because of inbreeding, and a lack of evolutionary potential. The loss of species has consequences for the web of nature and how it functions.

6 CHAPTER The web of nature

A MALE SCARAB BEETLE IS HEAD DOWN TO THE GROUND, with his back legs resting upon a sphere of dung that he is rolling along backwards, like a fitness fanatic trying to exercise with an out-of-control Swedish ball. A female scarab is standing by, observing proceedings as though she were a personal trainer, but her interest in what is going on is greater than would be normal in any dispassionate instructor. When the ball reaches soft ground, both beetles excavate and bury the dung underground. The beetles then mate and the female lays her eggs in the ball. After the eggs hatch, the dung provides the young larvae with their first meals.

From this small drama can be unfolded an entire ecological food web. We have already tangled with this web at breakfast (see chapter 1). The cow that provided the milk and yoghurt for breakfast is a herbivore – she eats plants. The scarabs and their offspring dine on what is left when the cow's digestive system and its community of intestinal helpers have extracted what they can from the food. Processing waste is a vital service provided by these beetles and their fellow decomposers (of which more later). When cattle were first introduced to Australia, the indigenous dung beetles were unable to dispose of such large dung pats because the continent has no native large ruminant mammals. Unburied cowpats provided breeding places for flies or baked hard in the sun and lay around the outback, preventing grass growth. The solution to this continental-scale constipation blocking the passage of dung through nature's web was to introduce bigger species of dung beetles from Europe and Africa. Dung burial by the introduced beetles not only reduced breeding places for flies and parasites, but increased pasture production through the nutrients that were returned to the soil where plants could reuse them.

The scarab beetle was sacred in ancient Egypt and its image appears in tomb paintings where it is associated with the solar deity Khepri. The sacred scarab, moving its ball of dung that it then buries and from which new scarabs eventually emerge, represented to the ancient Egyptians the sun moving across the sky, to be buried beneath the Earth at sunset and reborn each morning. Sun worship was not unique to Egypt and was practiced in ancient times from pre-Christian Europe to Aztec Mexico. All life is, after all, directly or indirectly solar-powered.

A solar-powered web

Plants capture the sun's energy in the process called photosynthesis. They use the power of the sun to turn carbon dioxide (CO_2) from the air and water from the soil into the sugar glucose that with other molecules make leaves, wood and other tissues. When

OPPOSITE The scarab beetle seen here in ancient Egyptian carving had a sacred association with the solar deity Khepri.

BELOW Burning releases the energy of sunlight that is locked up in wood.

you toss a piece of wood or paper onto a fire and it bursts into flame, you are releasing the solar energy captured by trees and locked away in their wood. Burning liberates not only the energy captured by photosynthesis, but also produces CO_2 and water vapour. Living things obtain energy from the products of photosynthesis by a process that is chemically equivalent to burning called respiration. Obviously, respiration does not produce flames because it is carefully controlled by enzymes that operate at lower temperatures. However, just as in a fire, the energy in the fuel we call food is ultimately liberated as heat (think of how your body stays warm) and as you would expect, respiration also produces CO_2 and water. You can see the water vapour condensing from your breath on a cold morning.

The pathway of carbon in its many guises through the web of life, including its embodiment in CO_2 in the atmosphere, and in wood, leaves, beefsteak, milk and dung, is known as the carbon cycle. Because carbon compounds make the fuel and the fabric of living things, the activity of organisms is crucial to how carbon flows around its cycle, how much accumulates and where, and how much is liberated into, or removed from, the atmosphere. So great is the impact of growth by trees in northern temperate forests on the atmospheric concentration of CO_2 that the annual cycle of summer growth and winter dormancy is visible as regular fluctuations in the CO_2 recorded on a mountaintop in Hawaii. The measurements summarized in the graph below also show that CO_2 in the atmosphere has been inexorably rising over the last 50 years (see opposite).

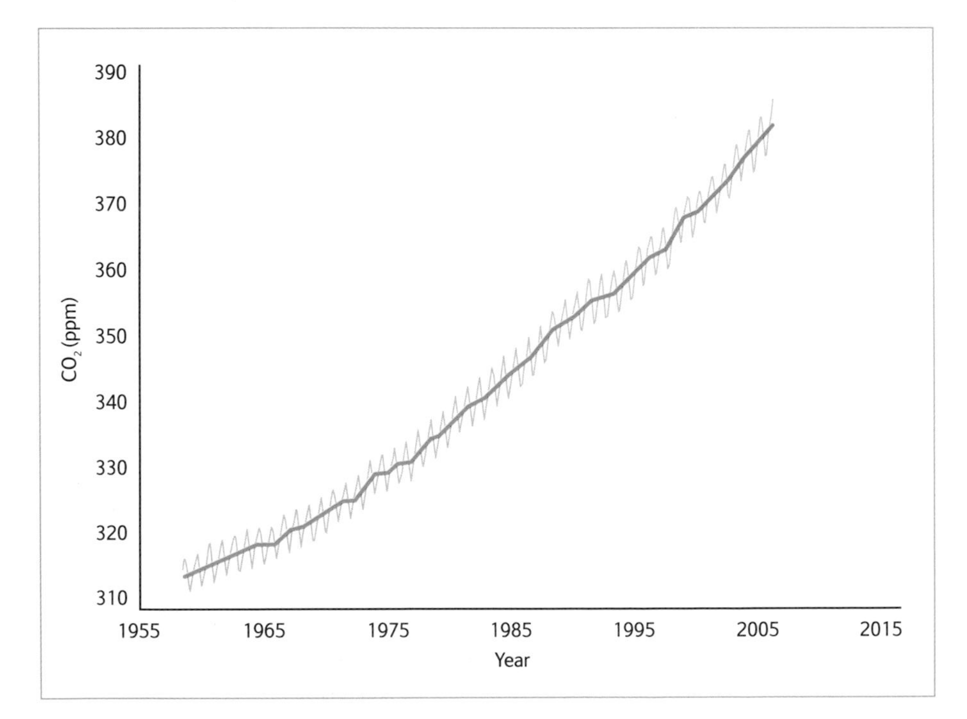

RIGHT A remorseless rise in the amount of CO_2 in the atmosphere has taken place, caused by emissions from burning fossil fuels and from deforestation. These measurements made at Mauna Loa, Hawaii, also show the annual cycle in atmospheric CO_2 concentration caused by the seasonal growth of trees in the northern hemisphere.

CLIMATE CHANGE AND THE CARBON CYCLE

Coal, gas and oil are the carbon residues of photosynthesis that occurred millions of years ago. Burning these fossil fuels adds more CO_2 to the atmosphere than living land plants and phytoplankton in today's world are capable of re-absorbing. As a result, CO_2 has accumulated, rising 50% above the concentration that was present in the atmosphere before the Industrial Revolution began. CO_2 contributes to the insulating properties of the thin gaseous blanket around our planet. The big increase in CO_2 concentration from burning fossil fuels is the principal cause of global warming.

BELOW The effect of various factors that influence atmospheric greenhouse properties. Human-related factors, and CO_2 concentration in particular, have the strongest effect and benefit from a high level of scientific understanding (LOSU).

	Factors	Cooling	Warming	LOSU
anthropogenic	long-lived greenhouse gases		CO_2	high
			N_2O, CH_4, halocarbons	high
	ozone	stratospheric	tropospheric	med
	surface albedo	land use	black carbon on snow	med-low
	total aerosol – direct effect			med-low
	total aerosol – cloud albedo effect			low
	stratospheric water vapour from CH_4			low
	linear contrails			low
natural	solar irradiance			low
	Total net anthropogenic			

–2 –1 0 1 2 3

radiative forcing/W m^{-2}

Biodiversity in the soil

Dung and the other detritus of life fall at the meeting-place between an above-ground world of familiar plants and animals and a netherworld of amazing, hidden biodiversity. Through the food web that joins these two worlds pass the elemental materials that life needs to function – nitrogen (N), phosphorus (P), potassium (K) and several other elements needed in smaller but crucial amounts. The carbon compounds in dead matter provide an energy source for the organisms that feed on it. The energy locked up in a pile of decaying manure or a compost heap is released as heat by the microbial decomposers, generating noticeable warming that, under the right conditions, can ignite the material.

RIGHT Steam rises from a compost heap due to the high temperatures generated by decomposition.

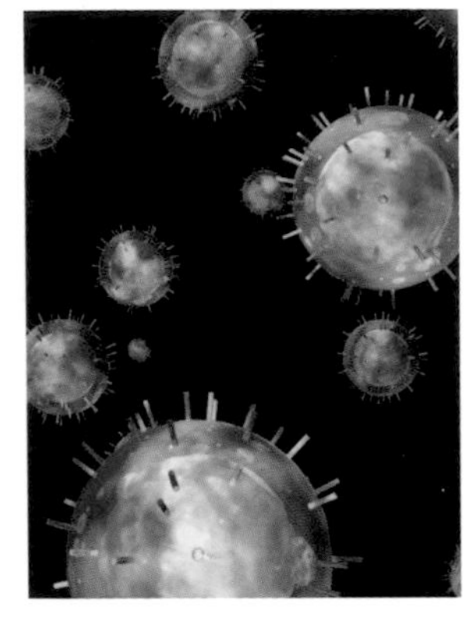

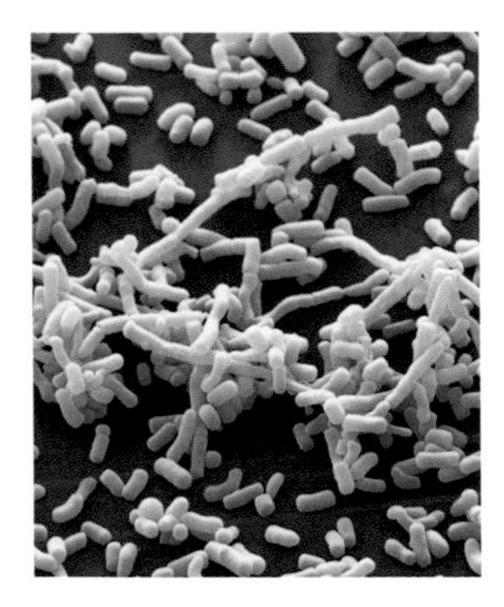

TOP Artist's impression of influenza virus particles. Viruses use the genetic machinery of a host's cells to build more copies of themselves, destroying the host's cells in the process.

ABOVE Coloured scanning electron micrograph (SEM) of a species of *Streptomyces* bacterium. These soil bacteria are responsible for the musty odour of soil. The magnification is x1,600.

Biodiversity below ground is even richer than above ground, but the organisms are mostly much smaller and therefore harder to count. About 10,000 species of bacteria have been scientifically named and, although not all can live in soil, there are two reasons to believe that this number does not nearly do justice to the actual biodiversity of soil bacteria. First, most descriptions of bacteria depend upon isolating them and then characterizing them by growing them in culture in the laboratory. The vast majority of soil bacteria have very specific, largely unknown, growth requirements and cannot be cultured, so they have not been conventionally named. The second issue is that bacteria do not comfortably fit the concept of 'species' that is used for eukaryotes (*see* chapter 4). Bacteria are infected by numerous viruses many of which quite promiscuously grab large amounts of DNA from one bacterium and insert it into another, creating the kind of identity crisis that you would have if you did not know which species of primate your father belonged to. Viruses are the ultimate parasites, stripped down to just a small length of DNA (or sometimes, as in the human immunodeficiency virus, the related molecule RNA) containing a handful of genes, protected inside a coat of protein. Viruses inject their DNA into a host cell where the cellular machinery normally used to make more host cells is hijacked to produce virus particles instead. Within a short time of infection, the host cell bursts and liberates new virus particles.

Fortunately, there are other ways of measuring the biodiversity of microbes in soil than trying to pigeonhole them into species. One way is to extract DNA from a bulked sample of all the microbes in the soil and then to compare sequences for a specific, slowly evolving gene within the sample to see how different the various copies are. Two sequences are defined as belonging to the same type of organism if they are 97% the same. Using this degree of similarity as a yardstick, one can count how many distinct types there are. The distinct types are *not* species, but they can be thought of as approximately that. One study using this technique found that in the top 5 cm (2 in) of soil there were thousands to tens-of-thousands of bacterial types and similar numbers again of archaeans, of fungi and of viruses in a plot only 10 m by 10 m (about 11 yards by 11 yards). When soils from desert, prairie and rainforest were compared

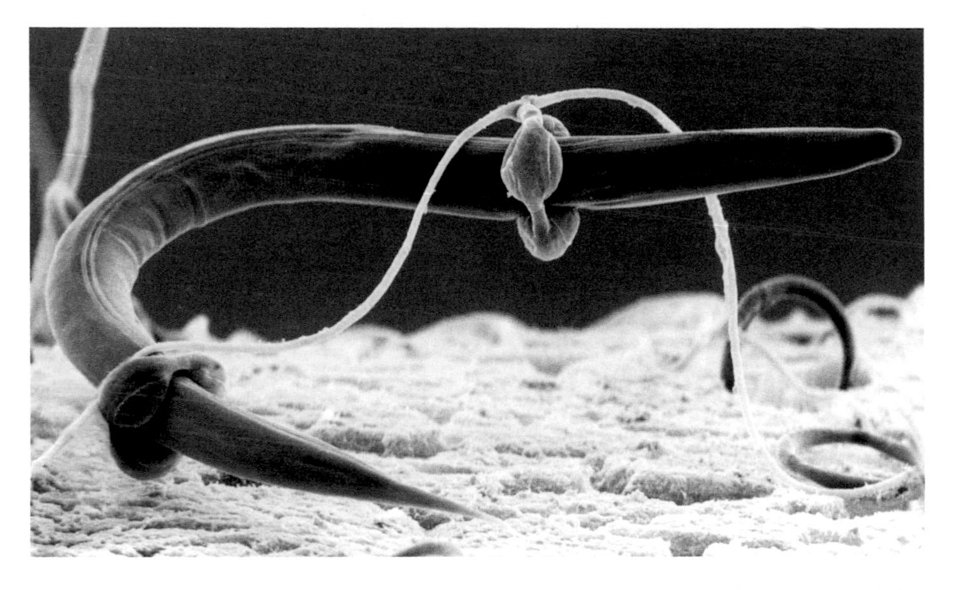

LEFT A nematode worm trapped in a tiny snare made by a predatory fungus that lives in soil. Width of field is about 200 microns or ⅕ of a millimetre.

with one another, desert soil was especially rich in archaeans; rainforest soil had the highest numbers of bacteria (and viruses) and prairie soil was richest in fungi. Not only was there a huge biodiversity of each of the microbial groups within the three kinds of environment, but each also harboured different bacteria, archaeans, fungi and viruses from each other. Together they add up to an awful lot of biodiversity in even a teaspoonful of soil, but we are not done yet!

There are several other very numerous groups of larger organisms in soil: amoebae and other one-celled organisms, and animals such as mites, nematode worms, springtails, ants, termites, beetles and, of course, earthworms, to name only the most important ones. Identifying and counting the numbers of species in these groups is easier than for microbes, but it is still not simple. For example, a genetic study of British earthworms found that individuals that looked alike were genetically distinct enough to be considered different species and that the nine described species in the study could conceal another seven cryptic ones. In fact cryptic species, identified by differences in their gene sequences, are showing up among many animals and plants (*see* chapter 4).

What are all these species doing in the soil? All play a part in the food web and they either feed on living plant roots, like certain nematode worms (e.g. those that attack potatoes) or soil fungi or bacteria, or they eat dead plant material, consume the faeces of other animals (think of the scarab) or they are predators. Every one of these ways of life is found among the hundreds of species of nematode worms typically found in soils. Mites (distantly related to spiders) and fungi are two other large groups that have a wide range of ecological roles. There are even predatory fungi that set snare-like traps for nematodes and help to control those that are pests of economically important plants.

Many of the fungi in soil decompose dead plants, others like the honey fungus (*Armillaria mellea*) are pathogens and attack living plants. Honey fungus can live also on dead wood and an ability to combine different sources of nutrition seems common among fungi. The mycorrhizal fungi are a very important group that form intimate associations with living roots and provide plants with access to phosphorus-containing minerals in exchange for glucose from the plant. Such reciprocal relationships that benefit both partners, called mutualisms, are not uncommon in nature. (The relationship between plants and the insects that pollinate their flowers is another example.) Some mycorrhizal fungi also live on dead matter. Nearly all plant species rely upon mycorrhizal fungi, which can also protect them from drought and from attack by nematodes. A few plants, such as certain orchids, are even parasitic upon fungi, taking energy-releasing compounds as well as minerals from the fungus.

Earthworms and termites eat dead plant material, but are also important in the physical effect that they have upon the soil, turning it over and creating a crumbly texture that improves aeration and drainage. Burrowing earthworms can ingest up to two and half times their weight in soil a day, amounting to tonnes of soil per hectare per year.

RIGHT Fungi are important in the decomposition of organic matter such as dead wood. Many are also pathogens of plants and animals.

RIGHT Earthworms are important to the fertility of soil, they eat dead plant material and earthworm activity aerates and mixes the soil, helping to improve drainage.

The organisms in a food web and their physical environment are collectively described as an ecosystem. An ecosystem is more-or-less self-contained, but may be of any size, from a garden pot containing a plant and some soil to an entire forest or an ocean. Energy and materials move through an ecosystem, largely in the bodies or through the activities of the organisms that live there. Consequently, each species has a function in the ecosystem, such as providing energy through photosynthesis or decomposition (many soil organisms) that releases essential nutrients from dead matter, making them available to plants. Another important function in the ecosystem is performed by bacteria that live in association with the roots of plants that belong to the pea family. These bacteria turn atmospheric nitrogen gas into soluble compounds that plants can use (*see* chapter 1). These functions are found in every ecosystem, but which species provide them differs between ecosystems, as the example of oceans shows.

Oceans

Breath in and out a couple of times. Your second breath comes, ultimately, from the oceans where roughly half of all photosynthesis takes place. There are no trees or leaves in the ocean but the plants that on land remove CO_2 from the atmosphere and repay the debt with oxygen, have their counterparts in marine phytoplankton. The phytoplankton are a diverse collection of microbes that share one thing in common with each other and with land plants: they are capable of photosynthesis – an ability that they have acquired directly or indirectly from cyanobacteria (*see* chapter 2). In fact, cyanobacteria themselves account for about half the photosynthesis in the ocean, with much of the rest contributed by diatoms. There are perhaps 200,000 diatom species, each a variation on a common pattern: a tiny glass box of two halves, with a photosynthetic cell inside.

A cubic metre of sea water contains trillions (millions of millions) of microbes, so many and so small that the only way to describe and count them comprehensively is to treat sea water as a DNA soup and to analyse the DNA sequences of the microbes in it. The sequences give a clue as to how many different kinds of micro-organisms there are and what roles they play in the marine environment. One result to emerge from such analyses is that viruses outnumber bacteria by about 10:1.

Just as on land, the whole ecosystem of oceanic surface waters depends upon solar energy captured through photosynthesis. Phytoplankton are eaten by tiny animals called zooplankton, that in turn are eaten by larger animals and so on up the food chain to top predators such as tuna or sharks. While life is powered ulitmately by the sun both on land and in the ocean, how the energy is channelled through the ecosystem and what happens to carbon are very different between land and sea. The reason is the contrasting sizes of the organisms doing the photosynthesis. On land, most plants are large and long-lived, while in the sea the tiny phytoplankton have the briefest of lives, often curtailed by being eaten or by a virus infection. The carbon that a land plant stores in its wood is removed from the atmosphere for many years, perhaps even millennia, in the form of peat and eventually coal if conditions prevent dead plant material from decomposing. In the ocean, phytoplankton multiply very quickly and can form massive populations, called blooms, in a matter of days, but they die quickly and shed the carbon that they have captured in their cells back into the ocean. Here it may be decomposed and turned back into CO_2 gas or the dead cells may sink to the bottom where the carbon is buried in ocean sediment. Which of these two fates occurs – release or burial – affects the carbon cycle. Burial of the dead phytoplankton in sediments represents a draw-down of carbon dioxide from the atmosphere. From this fact came the idea that phytoplankton blooms might be used to lower the concentration of CO_2 in the atmosphere as a remedy for global warming.

The idea of using artificial phytoplankton blooms to remove CO_2 from the atmosphere is an example of geo-engineering – or using a planet-sized technical fix to solve a global environmental problem. Diatoms, which contribute about 20% of global photosynthesis, require a number of raw materials for growth, principally nitrate, silicic acid (for their tiny glass box of a shell made of silica), phosphate, CO_2 and iron. Usable iron is produced by rock weathering and blown as dust into the sea. In areas of oceans that are very remote from land, it is iron that is in shortest supply and so it is this element that limits diatom growth. Therefore, fertilizing the ocean with iron particles should create a bloom of diatoms, which should remove dissolved CO_2 from surface waters. The big question is – what would happen next? Would the extra carbon captured by the diatom bloom be buried or would it be consumed by other organisms and re-released to the atmosphere? When the experiment was tried by adding iron to a large area in the northeast Pacific Ocean, the expected diatom bloom was generated and more CO_2 was indeed taken up, but only temporarily. The extra carbon in the diatom bloom was mostly consumed in the zooplankton or decomposed and

RIGHT False-colour satellite image of a phytoplankton bloom resulting from an iron fertilization experiment in the Gulf of Alaska. The location of the experimental bloom is visible as an orange patch surrounded by a fertilized blue ocean at the bottom centre of the image.

re-released. The additional amount sinking into the deep ocean was small. It has been argued that these experiments should be discontinued because even if the technique did remove additional carbon to the ocean depths, it could have huge, unpredictable and potentially disastrous consequences for ocean life, including depleting oxygen and essential nutrients.

Although fertilizing the ocean with iron turned out to be too simplistic a solution to global warming, the life in the ocean does provide us with plenty of other services, particularly food.

CHAPTER 7 *What has biodiversity ever done for us?*

THOUGH IT IS NOT A VERY PICTURESQUE SIMILE, ecosystems are like wholesalers for goods and services. When we fill our baskets with the weekly shopping at our local store, all the food we buy has come from the ecosystems that supply that shop in wholesale quantities. Only the fact that most of us do not grow our own food, but instead buy it neatly packaged and branded, obscures this obvious reality. If you have to grow a crop for yourself, knowledge of what role different organisms play in the ecosystem is invaluable. For example, introducing earthworms to the previously depauperate soils of a tea plantation in India increased yields by over 200%, thereby adding thousands of dollars per hectare to the profit obtained from the crop. Inoculating soils with the best-performing types of mycorrhizal fungi can dramatically improve the growth of many crops. The bacterial nitrogen-fixers attached to the roots of clover and related plants provide a source of nitrogen in pastures. In rice paddies, a floating aquatic fern called *Azolla* harbours cyanobacteria that can contribute 76 kg nitrogen per hectare per year (68 lb per acre per year) to the ecosystem. Such natural sources of usable nitrogen obviate the need to add nitrogen fertilizer, which is expensive to buy and energy-costly to make and to transport. All of these useful organisms, plus the insects that pollinate crops and many more can be found in the ecosystem supermarket next to the food section, in a well-stocked aisle labelled *Farmers' Friends*.

OPPOSITE Fish are the only major food that we still obtain from wild populations, but all major fisheries are over-exploited.

ABOVE *Azolla*, an aquatic fern used in rice paddies. This natural source of usable nitrogen obviates the need to add an artificial nitrogen fertilizer.

LEFT A tea plantation in the Himalayas, India. Earthworms in the soil help to increase production.

At the far end of the *Farmers' Friends* aisle, hidden discretely from view to spare the blushes of customers who have abused their ecosystems and are suffering the ugly consequences, is the *Ecosystem Reclamation* section. For example, a remedy for ecosystem degradation in the Sahel can be found in this section.. In this semi-desert region of Africa live some of the poorest people on Earth, most of them engaged in subsistence agriculture. Continuous cultivation and grazing and trampling by livestock combined with drought have caused soils in the Sahel to become covered in a crust that rain cannot penetrate and that is bare of vegetation. A simple remedy is to cover the surface in a mulch of straw and twigs, which is eaten and incorporated into the soil by termites that are already present in the area. Termite activity breaks the soil crust and improves soil structure, allowing rain to percolate downwards and enabling plants to grow. After 3 years of such a treatment an experiment in Burkina Faso found that treated plots produced over 3 tonnes per hectare (1.2 ton per acre) of plant dry matter, while untreated plots produced nothing at all. Plots with mulch but no termites produced only one-third the plant yield that mulch plus termites produced.

Water Supply occupies another aisle in the Ecosystem Services supermarket. Instead of the pipes and plumbers' fittings you might expect to find on the shelves, there are trees. But read the instructions before use because a tree in the wrong place can have the opposite effect to that intended. In most environments, forests protect water catchments, evening-out the flow and preventing soil from entering streams and polluting water supplies. New York City obtains most of the water for its 8 million inhabitants from a rural area of about 5,180 km^2 (2,000 $miles^2$) in the Catskill Mountains. The forest ecosystems in the Catskills naturally filter the water that runs through them, providing clean, drinkable water that does not require expensive filtration. The water is so good that it regularly beats expensive bottled water in tasting trials. To preserve this valuable ecosystem service, the city of New York has a long-running programme of buying-up land in the Catskills to prevent developments that could pollute its water supply. If the water supply became polluted, it would cost more than $6 billion to construct filtration plants to clean it up. Running the plants and servicing the $6 billion debt would cost at least a billion dollars a year more. These figures give some idea of the monetary value of clean water as an ecosystem service.

In the Mediterranean-type climate of the Cape in South Africa, the natural vegetation is not forest but a type of shrubland called 'fynbos', which is very rich in endemic plant species. The mountain catchments that feed the water reservoirs of Cape Town are covered in fynbos plants that use little water themselves. However, non-native trees are invading many fynbos habitats. These newcomers not only threaten the native flora, but can also reduce water supplies because they take up much more water than the native fynbos plants. A very successful campaign called Working for Water has been run in the Cape to remove non-native trees – simultaneously protecting water supplies and biodiversity and providing many local jobs.

Whether trees improve or reduce the flow of water from a catchment depends upon the balance between the effect of tree roots in increasing the water-holding capacity of soil on the one hand, and the amount of water used by trees and returned to the atmosphere through their leaves on the other. A further complication is that in certain environments, including in the Amazon rainforest, trees alter the local climate and removing them reduces rainfall. The lesson from all of these examples has to be that removing native vegetation can be a costly mistake.

Waste Processing is a service provided by living organisms too. Domestic waste water is treated biologically in sewage plants by filtering it through beds of gravel that are colonized by bacteria that remove nutrients, making the water safe enough to release back into rivers. Excess nutrients in lakes and rivers promote the growth of algae that then die at the end of the season. The bacteria that decompose the algal remains proliferate and deplete the water of oxygen in the process and as a result fish die in large numbers.

ABOVE Vultures provide an essential ecosystem service in India and their dramatic decline has probably cost thousands of human lives.

Cleaning up of ecosystems is not entirely a bacterial affair. In India, vultures were traditionally relied upon to consume the carcasses of dead cows and buffalo, but vultures declined very suddenly by 95% during the 1990s. The cause was at first very difficult to identify, but it has now been shown to be the result of the widespread use of a veterinary medicine given to livestock that is poisonous to birds. The repercussions of the vulture decline have been serious. A flock of vultures can reduce a cattle carcass to bare bones in a few hours, but with this service no longer available, feral dogs scavenged the dead cattle and with so much extra food available their numbers increased by about a third. Rabies, which is transmitted to other mammals including humans by bites from infected dogs, is endemic in India and it has been estimated that up to the year 2006, nearly 50,000 additional deaths from this dreaded disease may have resulted from the increase in dog numbers caused by the vulture decline.

LEFT Clouds above the rainforest in the Amazon. Up to half the rain that falls in the area is recycled through the trees, which are therefore vital to the local climate.

Before we leave the ecosystem supermarket, let us pass by the travel counter to see what holidays they have on offer. It is next to the pharmacy where plants that are the original source of so many useful drugs are kept. If you are planning a trip to a malarial area, you can collect some artemisinin tablets here, which is an effective remedy derived from the herb *Artemisia annua*. At the travel counter, see how prominently the vacation ads feature palm trees, forested landscapes, pristine lakes and quiet spots for fishing, flowers, coral reefs and fish, dolphins and other wild animals. Biodiversity makes a major contribution to recreation and to the types of environments where so many of us prefer to spend our leisure time. Income from ecotourism is vital to many countries rich in biodiversity. The majority of the foreign currency earned by Tanzania, for example, comes from tourists visiting the Serengeti and its other game parks. Tourism is the world's biggest industry and ecotourism has a growing share of the market. While the success of ecotourism illustrates yet another service provided by ecosystems, and biodiversity and its earning potential provides a commercial motive for preserving them, there is also the implicit threat of over-exploitation.

Resistance is useful

> *'I am Locutus of Borg. Resistance is futile. Your life as it has been is over. From this time forward, you will service us.'*

With these words, uttered in an episode of *Star Trek*, the captain of the *Starship Enterprise* learns that he is going to join the ranks of the exploited, providing services for another species. Our relationship to nature is not unlike that which the Borg intended for the captain and crew of the *Enterprise*. Given our dependence upon ecosystems and the species in them, how can we be sure that nature can carry this burden? How resistant to human impact are ecosystems? Contrary to the words of the Borg, when it comes to ecosystems, resistance is useful, not futile. A *resistant* ecosystem is one that does not change under the impact of exploitation. If it does change, we can ask how *resilient* is it? Resilience is the ability to bounce back after an impact. Resilience is useful too.

Resistance and resilience are properties of ecosystems, but their efficiency depends upon how living organisms respond to being exploited. Very moderate felling of trees in a forest should permit most of its ecosystem functions and services, such as regulation of water supply, to continue and hence the ecosystem could be said to be resistant to this level of exploitation. Clear-felling of patches of forest has a larger impact, but in resilient ecosystems the patches are eventually re-colonized by trees, and recover normal ecological functions.

In the nineteenth century, chestnut was an abundant and economically very important tree in the Appalachian forests of eastern USA, providing edible nuts, rot-resistant timber and bark for tanning leather. Chestnut blight, caused by an introduced fungus, wiped out the species in the first half of the twentieth century. The valuable products once obtained from chestnut were irretrievably lost, but other tree species like oaks and maples took the chestnut's place in the forest. Because the forest ecosystem

ABOVE Chestnut trees killed by blight, an introduced fungus that saw the demise of the species in eastern USA in the first half of the twentieth century.

as a whole recovered, it could be regarded as resilient in this respect, but the farmers living on the edge of subsistence for whom chestnut trees provided a valuable source of income might not agree.

How many of the species in a food web are needed to keep the ecosystem of which it is a part functioning? Is each species as important as every other, or can one substitute for the absence of another, as happened when feral dogs took the place of vultures in India? This question is about *redundancy*. How many species are irreplaceable in the role that they play in ecosystem function and how many are shadowed by a waiting line of alternatives that could take their place? High redundancy means that there are many alternatives. Low redundancy means that there are few alternatives. Redundancy can enable ecosystem functions to be sustained or to recover when a species is lost, and it can therefore contribute to resistance and resilience. Dogs may have taken over the scavenging functions of vultures in India, enabling decomposition (one ecosystem function) to continue, but not without a cost to human health (an ecosystem service). Hence, species performing similar ecosystem functions may be different from each other in important ways.

The food web of a savannah ecosystem such as the Serengeti or the Kruger National Parks in South Africa contains many species of plants, herbivores and carnivores. Some of the large carnivores are now extinct in the smaller African game parks and in such reserves, it has been found that cheetah numbers rise in the absence of lions and the endangered African wild dogs, *Lycaon pictus*, do better without the presence of

ABOVE Buffalo, one of the Big Five animals that draw tourists to the game parks of Africa.

spotted hyenas. These compensatory changes indicate that there is some redundancy in the food web. As in the case of vultures and feral dogs, or the American chestnut and oaks, such substitution can maintain some ecosystem functions, but others may still be adversely affected. Visitors to African game reserves are sold the idea that they should aim to see all of 'The Big Five' animals: lion, leopard, elephant, buffalo and rhino. There are no T-shirts for tourists who miss the lions that boast 'I saw only Four of the Big Five, but I did see some extra cheetahs'. The loss of any big species from such an ecosystem devalues its service to ecotourism. But, which species really count?

How many species do we need?

If you want to provision a passenger-carrying space ship for a one-way voyage of indefinite duration into the unknown, what must your onboard ecosystem contain? The bare essentials are photosynthetic organisms (plants or phytoplankton) and decomposers for treating waste and recycling essential elements. Animal consumers to turn plants into meat are optional, but including them makes designing a diet capable of sustaining the passengers much easier. A handful of species in each of these three categories might be sufficient to make the ecosystem function, but would you trust your life and that of future generations to such an impoverished sample of Earth's biodiversity? A stock of backup species would be a sensible precaution, but how many? Back on Earth, as our own numbers grow and large numbers of other species are threatened, we are de-provisioning our spaceship by jettisoning species with abandon. Species redundancy is an idea that may easily lull us into a false sense of security about the consequences of species loss. Oaks filled the gaps in the Appalachian forests left by the loss of the American chestnut, but now that the forest's stock is down by one major species, the scope for redundancy among the remaining species is also reduced. Oak wilt disease is now threatening those forests and chestnut is not there to take the oaks' place should the worst happen.

Research on several types of food web suggests that having many species makes them more resistant and/or resilient. For example, the majority of marine fisheries are under severe pressure from over-fishing and many fish stocks have collapsed to below 10% of their peak. Such collapses have been more frequent in areas where there were few fish species than in places where there were many. The causes are difficult to pin down, but where many species are available, fishers can more easily switch away from a species that is becoming rare to a more common one, allowing over-fished stocks to recover.

In grasslands, the presence of more plant species can increase the total productivity of the ecosystem and decrease its variation from year to year. Different species of plants have different periods of maximum growth during the year, have different good and bad years, or exploit different sources or kinds of nutrients. As a consequence, a species-rich community of plants can be better at making use of all available resources under all weather conditions than a species-poor one.

Higher plant productivity produced by increasing species diversity does not mean that adding species is the best way to achieve high yields (if that is the aim), because adding fertilizer or improving irrigation generally has a much greater effect. While adding resources (nutrients or water) increases total plant production, it may also reduce the number of plant species present because the strongest competitors for the added resources out-compete the weaker ones. For this reason the most productive natural ecosystems, such as estuarine marshes where nutrients and water are plentiful, are dominated by just one or two plant species. The types of natural vegetation with the highest biodiversity, such as calcareous grassland or 'fynbos' in the Cape, all occur on the most nutrient-poor soils. These associations show that it is impossible to achieve maximum plant productivity and high plant biodiversity simultaneously from the same ecosystem. From a practical standpoint, the relevant question becomes not so much 'How many species do I need?' as 'How many species do I want?' A choice, or at least a compromise, must be made between productivity and biodiversity.

All species are not equally important in ecosystems and as a broad generalization two things seem to determine which the important ones are: abundance and position in the food web. Unsurprisingly, abundant species are important in food webs. But, in virtually every natural community where species have been counted, there are only a handful of abundant species and the majority of species are rare. Is ecosystem function resistant to the loss of rare species? Possibly, but which species are abundant can change over time so that this season's bit-players might be next season's stars. To have no reserve of talent to cope with future eventualities is a bad idea, especially because there is growing evidence that ecosystems can flip between alternate states in which different species are important. In species-rich old grasslands such as intact prairie, for example, the decomposer food web in the soil tends to be dominated by fungi, while in fertilized agricultural grasslands, decomposition is dominated by bacteria and this

RIGHT Cod is a valuable fish that has been over-exploited in many fishing grounds.

difference has effects on the ecosystem. Because bacteria have short lives and rapid turnover, nutrients are much more available in the ecosystems where they dominate the soil food web and this abundance favours low plant species diversity. Pushing the food web back from a bacterially dominated food web to a fungus-dominated one may increase the plant biodiversity of a site, but it may be difficult to achieve.

Change in the abundance of large predators is one of the major causes of ecosystems flipping abruptly from one state to another. In the Baltic Sea in northern Europe a combination of fishing pressure and unfavourable environmental conditions for young cod caused the stock of this predatory fish to collapse in the 1980s. With its main predator drastically reduced, the sprat increased in numbers and the great abundance of this small fish depleted the zooplankton on which it feeds. Summer blooms of algae have become regular with the change to a sprat-dominated food web. Such blooms used to be prevented by zooplankton eating the algae, but phytoplankton-feeders are now less abundant because there are so many sprat feeding on them. Cod have failed to recover in the Baltic, possibly because their fry live among the zooplankton where sprat eat them.

Large predators at the very top of food webs (the so-called top predators) are much more important than their relatively low numbers would suggest. The reason for this is that so many other species are directly or indirectly linked to them through the food web. The American author and conservationist Aldo Leopold described over 60 years ago how he had seen the landscape change in response to the extermination of wolves:

> *‘I have lived to see state after state extirpate its wolves. I have watched the face of many a new wolfless mountain, and seen the south-facing slopes wrinkle with a maze of new deer trails. I have seen every edible bush and seedling browsed, first to anemic desuetude, and then to death. I have seen every edible tree defoliated to the height of a saddle horn.’*
>
> Aldo Leopold 1949

The reintroduction of wolves into Yellowstone National Park and other ecosystems in North America has revealed how astonishingly far-reaching the ecological effects of these predators may be. Ten years after reintroduction into Yellowstone, wolf numbers reached about 100 animals, but elk had declined by more than half from a peak of 20,000. Fear of wolves as well as predation itself had limited elk grazing, causing elk to stay away from areas of greatest risk near rivers where willows made it difficult to detect predators. The reduction of elk grazing in these areas allowed aspen trees to grow and to spread. In other places in Yellowstone where wolves are absent, elk grazing has prevented trees growing along river banks and led to erosion. Hence, operating through the food web, wolves are capable of indirectly remodelling the landscape. It is landscapes that we value.

BELOW Wolves can remodel a landscape through their indirect effect upon the vegetation eaten by their prey.

8 CHAPTER *Valued landscapes*

IN EVERYDAY LIFE, WHEN WE HAVE TO MEET OUR NEEDS for food and drink as efficiently as possible, many of us use the nearest supermarket, but when choosing a holiday we look harder for a destination that offers what we want. There is a huge range of options, often lavishly illustrated in brochures, but many of us choose destinations, like the mountains or the seaside, that bring us in closer contact with nature.

The fact that many of us now value nature, even in its wildest manifestations, stands in contrast with the situation at the start of industrialization. Then, after ten millennia of agricultural development, nature was seen as a threat to hard-won fields and pastures, a source of dangerous wild animals, weeds and pests. The wilder parts of nature, notably mountains and coasts were more a source of dread than of interest. Industrialization was built on a fusion of old and new attitudes. The new feature was the idea that nature was like a machine, so could be understood by science and controlled by technology. The old idea was that humans were separate from nature, which had been created by God for humans to exploit and dominate. This combination of values underpinned the industrialization and urbanization that began in the eighteenth century, achieved the wholesale transformation of the UK in the nineteenth century, and then subsequently spread ever more widely.

Against this mainstream, which still dominates, other values began to emerge, expressed both in the changing preferences for landscape and in a developing interest in natural history, leading to the development of ecology as a science and the recognition of the value of biodiversity.

OPPOSITE Yosemite, the world's first wilderness park. Initially seen as pristine nature, the vegetation of the valley floor was later found to have been shaped by indigenous people.

LEFT Ifugao rice terraces in the Philippines. A complex blend of nature and culture, sustained over 2,000 years but now at risk as young people move to the cities rather than remain as rice growers.

Eighteenth century anticipations

A significant innovation took place around 1725, epitomized by William Kent's design for Rousham Park in Oxfordshire, UK. Before this time, gardens had been fenced and were designed with obviously artificial forms, like straight paths, knot gardens and topiary. Kent was a leader in the development of what has come to be recognized as 'the English landscape garden', in which apparently natural forms of planting are highlighted by curved paths and carefully located follies or ruins, and the boundaries are hidden. In this way, the garden becomes an idealized landscape and seems to extend out into the surrounding countryside. The change in design, and of values, was paralleled by development of an international network of plant hunters and traders, with new plants eagerly sought in North America, South Africa and even further afield, and wealthy patrons paying high prices to obtain the first specimens for display in their gardens. Collecting live and preserved specimens also brought many animals into zoos and private collections, partly motivated by scientific enquiry, partly by public curiosity.

Later in the century, natural history also began to reshape appreciation of nature at both a local and global scale. The pioneer of detailed local study was Gilbert White, who spent decades studying the intricacies of his parish before publishing *The Natural History of Selborne* in 1789. A major international figure was the botanist Joseph Banks, who travelled as part of Captain Cook's first voyage to the Pacific Ocean from 1768 to 1771 and brought back specimens and records of hundreds of new species. He also helped to organize the collections at Kew Gardens, assisted by a colleague who had been trained by the taxonomist Linnaeus in Sweden, laying the foundations for its later role as one of the world's premier botanic gardens. In 1805, he was one of the founders of the Horticultural Society of London, later the Royal Horticultural Society, which was dedicated to the promotion of horticulture as both art and science.

The nineteenth century: Romanticism to ecology.

By the start of the new century a new movement was under way. Influenced by Jean-Jacques Rousseau's advocacy of seeking happiness in a natural existence, the Romantic Poets sought out wild nature and wrote poems expressing their emotional responses to mountains, storms and the admirable qualities of shepherds. Wordsworth's *Lyrical Ballads*, published in 1798, first brought the movement to a wide English-speaking public, but he was also known at the time as the author of an influential guide book which helped growing numbers of travellers to experience the landscapes of the English Lake District, which he described as 'a sort of national property', anticipating later campaigns to preserve it from development.

By the middle of the nineteenth century, as industrialization gathered pace and began to reach out across countries through railway building, a number of reactions to it were gathering strength. In New England, USA Henry David Thoreau was publishing books and articles advocating a simple life close to nature, best known through his novel *Walden; or, Life in the Woods*, published in 1845, but also travelling to wilder mountain landscapes,

LEFT The Chinese-style pagoda at Kew Gardens reflects the taste of the mid-eighteenth century for follies in a naturalistic landscape. Later in the century, Joseph Banks began the transformation of Kew into one of the world's premier botanic gardens and research centres, a World Heritage Site and major visitor attraction.

practicing natural history and advocating civil disobedience against pressures to conform to mainstream lifestyles. A complementary trend, intended to bring some of the benefits of contact with landscaped nature to city dwellers, was the beginning of the construction of urban parks, first in Birkenhead, UK in 1847, and soon after in Central Park, New York, then in hundreds of towns and cities. As news of the discovery of dramatically beautiful areas in the far west of the USA spread, the idea of public parks was also taken there, first in the establishment of Yosemite Valley and the Mariposa Grove of giant sequoias as a state park in 1864 and then of Yellowstone as the first National Park in 1872. The initial purpose of these parks was to promote tourism, initially as rail passengers seeking the most awesome views, and later through hiking and camping.

After 1859 when he published *On the Origin of Species*, Darwin's ideas began a profound change in understanding. He showed that the natural world was not machine-like but dynamic and creative and that humans were not created separately but evolved from, and in, nature. These ideas had the potential to challenge industrial exploitation, but Darwin's stress on competition was taken as a new justification of laissez-faire capitalism. Nevertheless, the idea of nature as dynamic and inter-related did contribute to the development of natural history into a new discipline, given a name in 1866 by Haeckel: ecology.

In 1864 another influential idea was launched though the publication of *Man and Nature* by a New Englander. George Perkins Marsh had travelled widely both in the USA and, as a diplomat, in the countries of the Mediterranean. Comparing descriptions in classical literature with what he observed on his travels, Marsh recognized that

Mediterranean landscapes had been degraded by human action, with former forests and rich agricultural land transformed into arid rock and scrub through soil erosion. Moreover, he also saw that deforestation was beginning to have the same effect in New England. Building on earlier recognition that islands could be degraded by deforestation and introduced animals, he demonstrated that human activity could transform whole landscapes and that care was needed to avoid land erosion.

All these trends came together in the life and work of John Muir, a Scot who emigrated to the USA with his parents in search of a better life. He took Thoreau's ideas of close contact with nature to the wildest parts of the mountains of California and recognized the damage still being done by sheep grazing and lumbering in Yosemite Valley and the Mariposa Grove in spite of their designation as a state park. Together with the editor of *Century* magazine, he mounted a campaign to alert the public to the wonders of the high Sierras and to press for more adequate protection, achieved through re-designation as a National Park in 1890. Muir's writing also spread ideas about the value of preserving nature to a wide public and laid the ground for the formation of the Sierra Club to promote the enjoyment and preservation of unspoiled nature. Both the designation of National Parks and the establishment of pressure groups to support preservationist causes spread internationally over the next century. Unfortunately, the US assumption that indigenous peoples should be removed from National Parks was often accepted as part of the package, especially where parks were designated by colonial regimes. An alternative idea was put forward by Ernest Thompson Seton, who argued that the 'American Indian' was a role model, and that city dwellers should learn woodcraft skills both to develop themselves and to learn respect for nature and supposed 'primitive' cultures.

In the UK, with its long settled landscapes and different cultural history, campaigns to preserve access to nature took different forms. One of the main features of the modernization of rural areas had been the enclosure of formerly common land by private landowners, depriving rural residents of traditional rights to graze animals or collect fuel, and sometimes also cutting off the footpaths they had used for local travel. In 1865 the Commons Preservation Society was set up to take action, both

BELOW The English Lake District is a glaciated landscape, transformed by agriculture and industry, celebrated by writers from Wordsworth to Beatrix Potter, and now a tourist destination and National Park.

direct and legal, to prevent unauthorized enclosure and in 1899 it merged with the Footpaths Preservation Society to form what is now known as the Open Spaces Society. The Lake District Defence Society, established in 1883 to campaign against inappropriate development in the Lake District, had a membership largely based in London. In 1895, members of a number of different campaigning groups joined together to set up the National Trust, a society dedicated to raising funds to gain ownership of valued landscapes and buildings to preserve them in perpetuity. In 2009 the Trust owns over 250,000 ha of land (617,500 acres), selected for historical, cultural and aesthetic reasons as well as for nature conservation.

The twentieth century: internationalization and biodiversity

The first half of the twentieth century, with its cycle of war, depression and further war, also saw further expansion of industrialization, at best tempered by Gifford Pinchot's interpretation of conservation as 'wise use' of natural resources for the public good, at worst driven solely by the profit motive. The National Park idea continued to spread, but there were few major innovations in attitudes in this period. In the UK, the Soil Association was founded to campaign for soil conservation and organic agriculture. The practice of woodcraft became widespread in the USA and Europe but the name faded after Seton's death in 1946, although many of his ideas live on through his influence on the Scout movement. Attitudes to nature were increasingly informed by scientific ecology, now becoming a recognized part of university biology departments, and with developing professional networks, for example the Ecological Society of America, set up in 1915.

In the late 1940s, as governments and people set about remaking the world in the hope that war and depression would not happen again, substantial steps forward took place. The UK government at last accepted the idea of designating National Parks in upland areas, though farms, villages and even small towns were included, so policy had to balance conservation and recreation with established economic activities. At the same time, the Nature Conservancy was set up with a narrower remit to preserve good examples of ecosystems in National Nature Reserves, usually with access only for scientists, and Sites of Special Scientific Interest, where existing landowners were encouraged to manage in favour of the special elements – although this was not always done, and on occasions designation seems to have reduced the site's value as agricultural land and encouraged development, including road building. In the USA, the Nature Conservancy was an offshoot of the Ecological Society of America, set up in 1951 as a charity to apply ecological expertise in the preservation and restoration of valued sites. By 1980, it had sites all over the USA and began to spread internationally.

Internationalization was a major feature of post-war conservation. The International Union for the Preservation of Nature was set up in 1948 as an umbrella organization to co-ordinate both government and non-government bodies. It was soon renamed the International Union for the Conservation of Nature and Natural Resources, usually known as IUCN, and campaigned for 'a just world that values and conserves nature', including promoting and monitoring 'protected areas', defined as 'an area of land and/ or sea especially dedicated to the protection and maintenance of biological diversity, and of natural and associated cultural resources, and managed through legal or other effective means'. The number and area of protected areas grew rapidly from the 1960s to the 1990s.

In the late 1960s and early 1970s, 'the environment' suddenly became an issue in local, national and international politics as many people began to accept that mainstream industrial and economic development was having detrimental effects. New pressure groups and government departments were formed in many countries and 'environmentalist' or 'ecological' values were taken more seriously. In this atmosphere the United Nations Educational Scientific and Cultural Organization (UNESCO) launched two major initiatives. The more familiar was its new system for recognizing World Heritage Sites, most on the basis of their cultural importance, but some for their natural importance. The 878 World Heritage Sites are a selection of world-class cultural and natural monuments, but UNESCO is stronger in its insistence on effective management than for any innovation in environmental values. The second initiative, the recognition of Biosphere Reserves, sought both to conserve valued ecosystems in a core area and to promote development in a surrounding area.

During the 1980s, scientists began using the term 'biodiversity' and to advocate conserving 'biodiversity hotspots'. This was taken up as part of the United Nations' promotion of sustainable development and a key aspect of the United Nations' Conference on Environment and Development held in Rio de Janeiro in 1992 was the agreement on the Convention on Biological Diversity. This advocated conservation of

RIGHT One of the world's most spectacular wildlife sites the Serengeti, Tanzania, was recognized as a National Park in 1951, and a World Heritage Site and Biosphere Reserve in 1981.

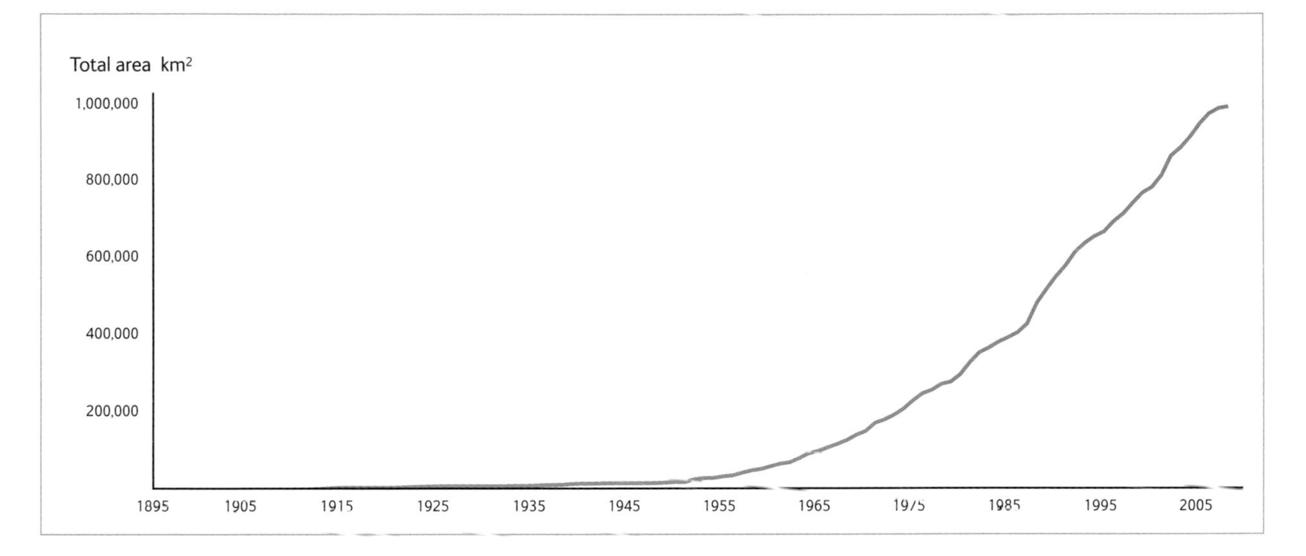

ABOVE The graph shows the growth in total area of nationally designated protected areas from 1872 to 2007.

biodiversity, its sustainable use and fair and equitable sharing of the benefits. In turn, these new ideas were used in the management and recognition of Biosphere Reserves. Whereas earlier nature reserves were often seen as preserving ecosystems from the past, Biosphere Reserves are seen by UNESCO as 'laboratories of the new', that is as pioneers of new ways of combining conservation and sustainable lifestyles, which could then be implemented more widely. In 2009 there are more than 550 Biosphere Reserves in over 100 countries.

One of the Commissions of IUCN, the World Commission on Protected Areas, maintains statistics on protected areas worldwide. Currently, there are more than 100,000 protected areas covering over 12% of the Earth's land surface, though less than 1% of the sea. However, this apparently enormous achievement is lessened by weaknesses in management and by the fragmented nature of any system of protected areas. The system is at best a fragile defence against growing threats from development and from climate change.

The twenty-first century: ecological disaster or restoration?

As the Millennium approached, many people and institutions felt the need to assess what had been achieved to date and what the prospects for the future were. Among them, the United Nations set up the Millennium Ecosystem Assessment, which documented the accelerating degradation of ecosystems. Three United Nations conferences on the environment had produced fine resolutions but had done little to affect the pursuit of economic growth at all costs. Worse, the fruits of growth were becoming even more unequal, leaving a third of the Earth's population in poverty. Boosted by an international campaign to 'make poverty history' the United Nations gained approval for a set of Millennium Development Goals to promote poverty reduction and sustainable development. However, although many countries signed up to the goals,

ABOVE A former chalk quarry, restored over 20 years to provide open water, marshland and chalk grassland habitats supporting rare species of birds, butterflies and plants.

there was no agreement on how they should be achieved, and progress has been slow. Instead, economic growth has remained the over-riding goal, with negative effects on biodiversity, as documented in many chapters of this book. Without policy change to seek a wider range of goals, the prospects for the future are bleak. Here we look briefly at a first step towards sustainable development – ecological restoration. Whereas existing protected areas are mainly a defence against development, ecological restoration uses the fact that industrial society has great power to alter landscapes and to restore areas that have been degraded by past use. Projects like species re-introduction, control of invasive species, creating wildlife corridors between reserves and even re-construction of whole landscapes are already happening, mainly in developed countries.

Perhaps the most ambitious and successful example of ecological restoration has been the Tallgrass Prairie Preserve in Kansas. A 4,050-ha (10,000-acre) reserve was purchased by the Nature Conservancy in 1989 and managed for bison grazing by patch-burning. This practice creates a varied mosaic of vegetation and the bison selectively graze grass and sedge, leaving wildflowers to bloom. Although a tiny fragment of a prairie environment that once covered 1,036,000 km^2 (400,000 miles^2) and supported 30 million bison before hunting reduced their numbers to only 1,000 in 1900, the Preserve has shown how a more beautiful and biologically-diverse habitat can be restored, and even become a model for grazing management by cattle ranchers. The Preserve was recognized as a National Park in 1996 and has been intensively researched by ecologists.

Although preferable to further extension of agricultural monoculture or urbanization, ecological restoration raises two sets of problems: technical and ethical. First, there are the technical problems of designing landscapes of particular types and then implementing that design, a process that needs to be based on increased knowledge of restoration ecology. Then there are the ethical problems, first of deciding exactly what the system is to be restored to: the Tallgrass Prairie National Preserve does not include wolves, the top predator of the natural prairies ecosystem (*see* chapter 7), and second, the potential problem of affluent societies taking land out of economic production to restore it for aesthetic or ecological purposes, and in consequence increasing their ecological footprint in less developed countries. These ethical issues are a reminder that sustaining biodiversity may be in conflict with social justice and are not easy to reconcile.

Sustainable development

Within the negative trend often stressed by environmentalists – population growth, compounded by economic and technological development and increased environmental degradation – are some more positive indications. Many human groups have had livelihoods that sustain, sometimes improve, their landscapes. The growing influence and diversity of ways of valuing nature that have emerged over the last three centuries suggest that richer societies tend towards non-materialist values, have improving knowledge of how best to manage for multiple use and even the ability to restore biodiverse ecosystems. Over a century's work on evolution and ecology have shown that humans are not separate from nature, which is dynamic and creative rather than machine-like. The cardinal assumptions behind industrialization are therefore false. The future need not be one of mass extinctions and lifeless oceans and cities, but to reverse the trend of exploitation and degradation will require a major change in human priorities. The original concept of 'sustainable development' included biodiversity conservation as well as equity between humans but it has not yet been seriously implemented. Perhaps the biggest challenge is to bring the conservation of biodiversity to the urban environments where more than half the human population now live. In fact there is more nature in cities than one might imagine.

9 CHAPTER *Nature in the city*

SINCE ANCIENT TIMES, CITIES HAVE BEEN REGARDED as the domain of humans and separate from nature. As the early cities became established in Mesopotamia and around the Mediterranean they were considered to be symbols of human dominion over nature. The removal of wild and dangerous animals from an area was considered to be part of the process of civilization, creating safe havens for humans and their domesticated animals. A relief found in the ruins of the ancient North African city of Cyrene shows the nymph Kurene (after whom Cyrene is named) killing a lion and being crowned by the goddess Libya. Demonstrating their control over nature the people of these ancient cities in arid regions used water lavishly in public baths, fountains and gardens. The ultimate example of this showmanship was the Hanging Gardens of Babylon.

The industrialization of cities in the eighteenth and nineteenth centuries brought with it pollution of the air and water. Cities became unpleasant places associated with filth and disease, unsuitable for humans, let alone wildlife. In the twentieth century, environmental legislation cleaned up the urban environment but improvements in transport such as railways and the motor car allowed cities to sprawl over the surrounding countryside as more people came to live in them. In 1900, 9% of the world's human population lived in cities. By 2000 this figure was 50% and by 2025 it is expected to be over 66%.

OPPOSITE The built environment itself is also exploited by wildlife. Kittiwakes, *Rissa tridactyla*, use buildings as surrogate cliffs for nesting, especially in coastal cities.

LEFT Industrialization of cities brought with it pollution making cities unpleasant and dangerous places.

Although the city environment is, in many ways, detrimental to wildlife, living organisms are surprisingly adaptable. Despite human attempts to tame nature and the effects of pollution, wildlife is present in cities. Ecologists have long regarded cities as artificial environments, so the study of urban ecology has been a neglected field compared with the study of more spectacular ecosystems such as tropical rainforests. How does wildlife exploit the urban environment? Is the global spread of urbanization always a threat to biodiversity?

Urban versus rural environments

Urbanization is not a single environmental parameter such as temperature or humidity. It is a complex combination of many factors including climate, pollution and human disturbance. The composition of the atmosphere changes gradually between the city and its rural surroundings. The differences between the urban and rural environments are most pronounced in the city centre and gradually become less pronounced towards the edge of the city. City centres tend to be warmer, less humid and less windy than surrounding rural areas. This characteristic urban climate is known as the urban heat island effect. One reason for this climatic difference between the city and the country is the properties of the materials used in buildings and for covering the ground surface. Brick, stone, concrete and tarmac absorb heat from the sun during the day and release it at night. Waste heat from the activities of humans (such as heating of buildings, air conditioning and motor vehicles) also contributes to the heat island effect. The covering of ground surfaces with impermeable materials, such as concrete and tarmac, diverts surface water into sewage systems rather than allowing it to be absorbed slowly by the soil. Less water evaporates from artificially drained surfaces, which not only reduces the humidity of the air, but also contributes to the heat island effect because there is less cooling by evaporation. The lack of vegetation cover in cities also reduces the potential for cooling by lowering rates of water loss through plant leaves.

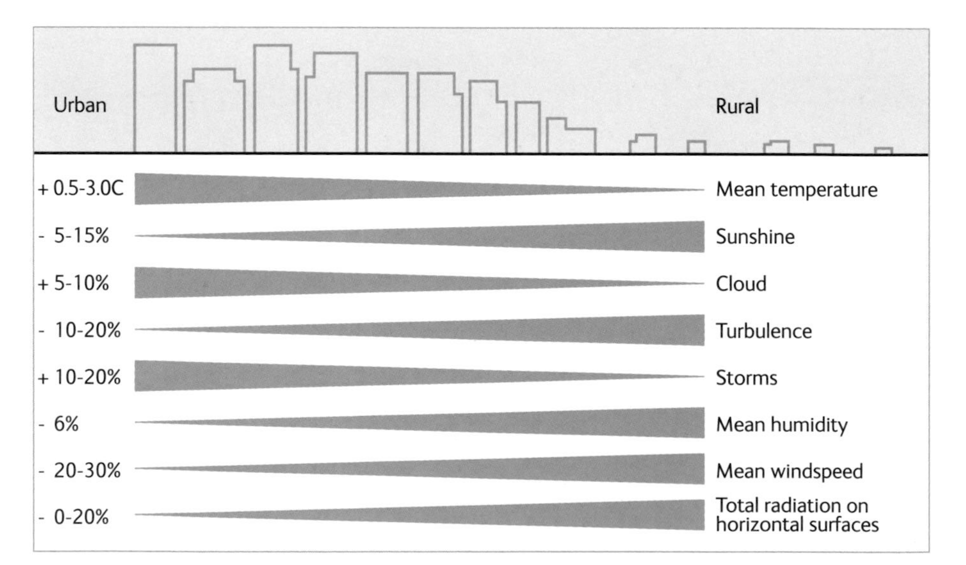

RIGHT A summary of the range of environmental characteristics which distinguish the climate of city areas from rural areas.

The detrimental effects of pollution resulting from human activities on the city environment have long been recognized and active attempts are taken to control and reduce pollution. However, when setting the minimum standards for acceptable pollution, legislators are more concerned with human health than with wildlife. Phosphate and nitrate levels in water that may be safely drunk by humans can seriously harm biodiversity in freshwater systems. In the UK the maximum acceptable level of phosphate in drinking water is 50 mg/l (50 parts per million) but biodiversity declines in water systems containing as little as 0.1 mg/l (0.1 parts per million). The use of salt to de-ice roads in winter is of no concern to human health but it can damage and even kill street trees in cities. In the UK, the saltmarsh plant *Puccinellia distans*, has spread from the coast to inland counties such as Derbyshire along the main roads, which are frequently salted in winter. Levels of sulphur dioxide pollution in the atmosphere that are tolerable to humans can be lethal to whole genera of the more sensitive lichens such as *Usnea* spp.

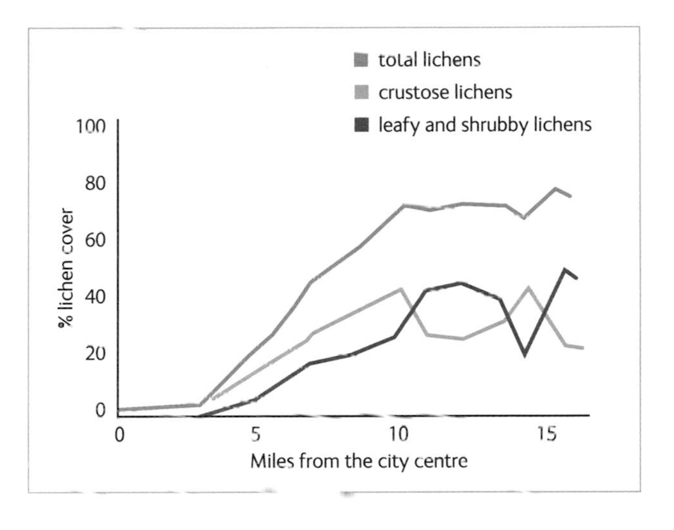

ABOVE The percentage cover of lichens on deciduous tree trunks from Belfast, Ireland.

A study of lichens along a sampling line (called a transect) from urban to rural environments in Belfast, Northern Ireland found that only crustose lichens were present in the city centre at very low abundance. A further 5 km (3 miles) out from the centre, the abundance of lichens was higher and species of the more pollution-sensitive shrubby and leafy lichens appeared. At 16 km (10 miles) from the centre, the crustose lichens declined and the leafy and shrubby species started to dominate. Other studies suggest that this distribution is the result of different tolerances and sensitivities of different lichen species to pollution by sulphur dioxide in the atmosphere rather than to climatic variables. The reduced diversity and abundance of lichens in city centres has an indirect effect on other organisms. A study in Newcastle, UK showed a decline in abundance of bark lice (Psocidae) in the city centre. These are small insects that feed on lichens on tree bark.

The city environment can have other indirect and sometimes unexpected effects on wildlife. Horse chestnut trees growing in the centre of Oxford, UK are more susceptible to infestations of a scale insect, *Pulvinaria regalis*, than those growing in rural areas. Soil compaction and impermeable ground surfaces, such as concrete and roads, weaken the city trees making them more susceptible to infestation. Urban holly bushes in Newark, USA have higher infestations of holly leaf miner, *Phytomyza ilicicola*, than rural bushes. The holly leaf miner is a fly whose larvae develop in tunnels which they bore inside holly leaves as they feed, making characteristic blotches on the leaves. Leaf miners are attacked by a parasitic wasp. A higher percentage of rural leaf miners are parasitized than those in urban areas, which could account for the higher leaf miner infestations in the city. An investigation into the reasons for the differences in abundance of leaf miner in rural and urban areas showed how the urban environment influenced the rate of parasitism. The rural bushes are naturally found in woodland so are more shaded than the urban bushes, which are planted in the open with no overhanging trees. The greater

RIGHT Holly leaf miner, *Phytomyza ilicicola*. Early leaf fall in cities allows the leaf miner to escape attack by a parasitic wasp.

exposure to sunlight in the urban setting leads to earlier senescence and dropping of the leaves. The leaf miners are still able to complete their development in the fallen leaves so are unaffected by the early leaf fall. However, the parasitic wasp does not search for hosts among the fallen leaves so many of the leaf miners in the city escape parasitism.

Where is the city wildlife?

Even large cities are not totally built up. There can be a surprising amount of open space in cities. In the UK, 35% of the land area in the major cities, including London, Birmingham, Liverpool and Manchester, is classified as green space.

The types of open space commonly found in cities include:

- remnants of countryside engulfed by the expanding city, such as old woodland, heathland, river valleys or farmland
- linear space such as road verges and land along railway lines and canals
- designed open space such as parks, sports fields, gardens, allotments and land associated with schools, hospitals, cemeteries and sewage works
- disused land, such as derelict industrial land, demolition sites and other disturbed waste ground (in the UK, derelict land is often referred to by planners as a brown-field site).

The amount of open space in cities is constantly changing. As cities expand into the countryside, pieces of natural habitat are engulfed and surrounded by built-up land. Such land is often preserved because it has characteristics, such as steep slopes, which make it unsuitable for building. Although it may seem fortunate that some natural habitat is preserved within cities, such fragmented habitat is often altered as a result of anthropogenic influences. A study of remnant heathland and woodland in Sydney, Australia, showed that assemblages of arthropods differed depending on the size of the fragment, especially for predators and parasitoids. Species richness was higher in small fragments because of the presence of a large proportion of generalist species at the expense of the rarer, habitat-specific, specialist species. It was suggested that the observed differences were the result of human influences. Small fragments were subject to more human disturbance (measured by the number of footpaths and the amount of human-produced rubbish present) than large fragments. Heathland is a habitat dependent on periodic burning, which prevents the development of scrub and woodland. Heathland fragments in the city are protected from fire, for safety reasons, so many urban heathlands become overgrown and subject to invasion by non-native shrub species.

Fragmentation can limit mobility between habitats. As a habitat deteriorates many organisms are unable to cross inhospitable areas to reach another suitable area. Linear open space such as railway embankments, canals (used or disused) and road verges provide relatively undisturbed habitat which can act as a wildlife network linking fragmented remnant habitats and urban areas with rural areas. They can also facilitate the spread of invasive non native species.

Even the largest cities contain areas of planned open space such as parks, cemeteries, churchyards and recreation grounds. Although these places are well tended they do provide habitat for wildlife such as trees and pond-dwellers. Cemeteries and churchyards

LEFT An aerial view of the Bermondsey area of London showing urban green space including playing fields, city parks, street trees, private gardens and undeveloped land along the edge of a railway. The Thames, Hyde Park, Regents Park and Battersea Park can be seen, in the distance, at the top of the picture.

are well known as places to find a wide range of lichen species on gravestones and walls. Gardens are important sites for insects and birds. Although individual gardens are small, the total area of UK gardens, many of which are interconnected, is over 270,000 ha (667,000 acres), more than the total area of National Nature Reserves. In a survey of gardens in Sheffield it was found that domestic gardens covered approximately 33 km^2 (13 $miles^2$), which is 23% of the urbanized area. In these gardens there were 25,000 ponds, 45,500 bird nesting boxes, 50,750 compost heaps and 360,000 trees more than 3 m (10 ft) tall. The number of ponds exceeded the density of ponds in the surrounding rural area. There were 1,166 vascular plant species (31% of which were native), 80 species of lichens (on average this was 15 per garden), 68 moss species (11 species per garden) and 700 species of insects.

The plants found in gardens are of great importance to urban wildlife. The range of nectar-producing plants present in gardens are an important food source for bumblebees and butterflies. As a result of intensive agriculture, the bumblebee populations are declining in the countryside so urban gardens are becoming even more important for them. The range of berry-producing plant species in gardens are an important winter food source for many birds, as is the human practice of providing supplementary food for birds. In the USA, a study of Florida scrub jays, *Aphelocoma coerulescens,* in suburban and rural areas found that supplementary feeding by humans in the suburban area improved the body condition of female suburban birds, enabling them to breed earlier in the year than the rural birds.

Derelict industrial land or vacant demolition sites (brown-field sites) have a characteristic flora consisting mainly of plant species that are adapted to colonizing recently disturbed soils, such as rosebay willowherb, *Chamerion angustifolium.* Many of these plant species are rich sources of nectar and are larval food plants. They are particularly important for butterflies and moths; indeed, in the UK there are 30 species of butterfly associated with derelict sites.

RIGHT The site of an abandoned oil refinery on Canvey Island in Essex has recently been declared a Site of Special Scientific Interest because it is so rich in insect species. Thirty Red Data Book species and three species previously thought to be extinct in Britain are known to occur there. Many orchid species are also present at the site.

The built environment itself is also exploited by wildlife. Gulls use buildings as surrogate cliffs for nesting, especially in coastal cities and that characteristically urban bird the feral pigeon, *Columba livia,* was bred by pigeon fanciers from the rock dove, which is a cliff-dweller. Bats exploit buildings as surrogate caves inhabiting crevices in roof spaces, cellars and tunnels. Sewage works provide habitat for a wide range of invertebrates that feed on the decaying organic matter in the sedimentation tanks and filter beds. They in turn attract a wide range of bird species to feed on them including crows, wagtails, swifts, martins and swallows.

LEFT Rosebay willowherb, *Chamerion angustifolium*, a species often found in areas of derelict land in towns and cities.

The nature of the urban threat to biodiversity

The discussion up to now suggests that some of the environmental differences between the city and the country, such as the warmer environment found in cities, are not necessarily detrimental and, to some species, may even be advantageous. So if some species thrive in the city environment and there is so much green space in cities, is urbanization really a threat to biodiversity? To answer this question it is necessary to look more closely at how biodiversity changes as countryside is replaced by city.

A long-term study of the flora of Berlin, Germany identified one of the threats to biodiversity of plants in the city as habitat is lost to urban sprawl. Historical records of ferns and flowering plants recorded in the Berlin area between 1859 and 1959 showed that over this period, 114 species (12% of the native flora) disappeared from the area. Particular groups of plants suffered more that others: 41% of pond weed (potamogetons) species were lost, as were many species of orchid. He concluded that the main reason for the loss of so many species was the loss of specialized habitats such as bogs, freshwater, agricultural fields, acidic grassland, moist meadows and woodland.

Records of the number of species of flowering plants made along a transect between the countryside and the city found that there was a sharp increase in the number of species per km^2 at the city outskirts. From the outskirts to the city centre there was a slight decline but species richness in the city centre was substantially higher than in the surrounding countryside.

Both native and non-native species were recorded along the gradient. In the countryside surrounding Berlin 28.5% of the species recorded were non-native, but fully half the species in the city centre were not native and these were responsible for the higher diversity there. In the city, many of the non-native species were found in disturbed places such as industrial and railway sites. There were more native species at sites with intermediate levels of disturbance and fewest species at sites with high levels of disturbance. On the other hand, there were more non-native species at sites with high disturbance levels.

BELOW LEFT Species richness of flowering plants along a rural to urban transect in Berlin. Zone 1 – surrounding countryside, Zone 2 – outskirts, Zone 3 – loosely built up, Zone 4 – city centre (closely built up).

BELOW RIGHT The relationship between species richness and the level of human disturbance in the city of Berlin. Disturbance is scored on a scale from one to nine in which nine is the highest level of disturbance.

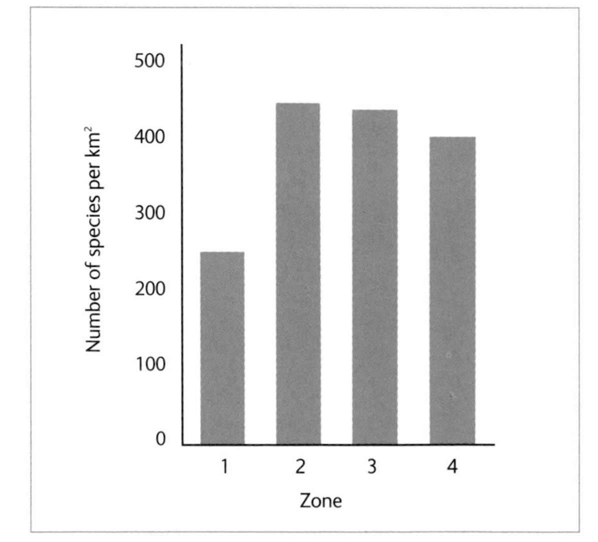

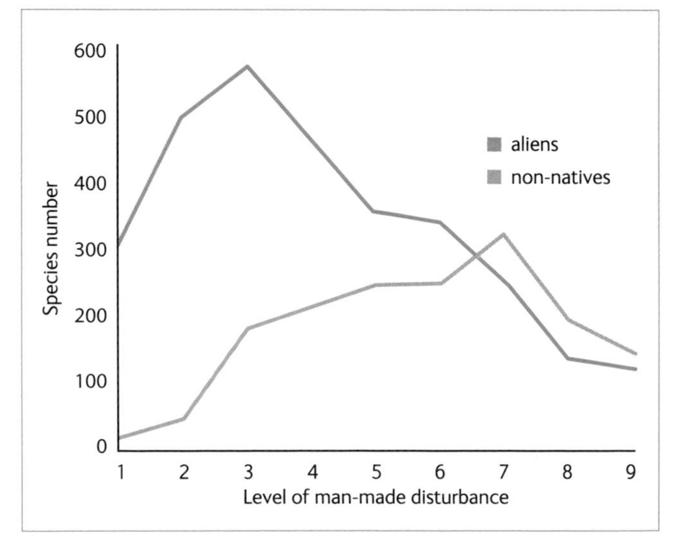

Cities are constantly changing. Vegetation is cleared to make way for new buildings but new opportunities for wildlife arise as buildings and land fall into disuse. The species likely to thrive in such an environment are invasive species that disperse readily and reproduce quickly. Many such species are non-natives introduced intentionally or unintentionally by the activities of humans.

In New York State it has been found that non-native species have an important ecosystem function in urban forests. Urban forest soils contain high levels of heavy metals that are toxic to most soil invertebrates and fungi. This might be expected to lead to slower litter decomposition and lower rates of nitrogen release from soils in urban forests, but a study found that on the contrary, the rate of litter decomposition was faster in urban forests than rural ones for two reasons. The urban heat island effect increased the temperature of soils, which speeded up decomposition, and the presence of high numbers of non-native species of earthworm increased litter decomposition rates. Both were the result of the activities of humans.

Making the most of nature in cities

As cities spread, native habitat is lost. Although cities contain a high proportion of green space that can be effectively exploited by wildlife, the species that thrive in the urban environment tend to be invasive species that readily disperse and reproduce quickly. Many of these species are non-natives, introduced, either intentionally or unintentionally, by the activities of humans. Based on this evidence, urbanization is a threat to biodiversity. However, the varied range of habitats present in cities means that many species are supported, many more than would be found in equivalent areas of intensive agriculture. Considering the growth of cities throughout the world and their ecological importance, more resources should be directed towards understanding the ecology of cities.

10 Life in freshwater

CHAPTER

'We're treating the planet as a toilet and we think when we flush the chain the problems disappear. They don't.' Craig Venter

OPPOSITE Freshwater, stored and filtered by nature, is one of the services that ecosystems provide, as well as being a habitat for many species.

ALL TERRESTRIAL ANIMALS AND PLANTS NEED FRESHWATER to sustain their lives. They need it, moreover, to be pure and free of chemical pollutants. Clean freshwater is a scarce resource in the natural world and, in the rapidly changing world that humans have created, it is becoming scarcer. Humans too need clean freshwater, not only with minimal levels of pollutants but also low numbers of the many kinds of microbes that cause disease. The number of people across the world who are denied access to a sufficient supply of clean, safe water is increasing rapidly. There is a global crisis in relation to freshwater that poses a major challenge to those who seek to conserve biodiversity – how to meet the needs both of wild animals and plants and of people?

Global freshwater resources

Seen from space the Earth is blue; it is a very wet planet. However, most of its water is in the oceans; only 2.53% of its water is fresh and freshwater covers less than 1% of the Earth's surface. Only a tiny proportion of freshwater, 0.1%, is immediately available for plants, animals and people, as lakes, rivers, streams, ponds and wetlands. The rest is 'locked away', 75% as ice in the two polar ice-caps and in glaciers, 24.99% deep underground in rock. Not only is available freshwater in short supply, it is also very unevenly distributed across the globe. For example, Canada has 30 times as much freshwater available to each of its citizens as China. Asia supports 60% of the world's human population, but has only 36% of the world's available freshwater.

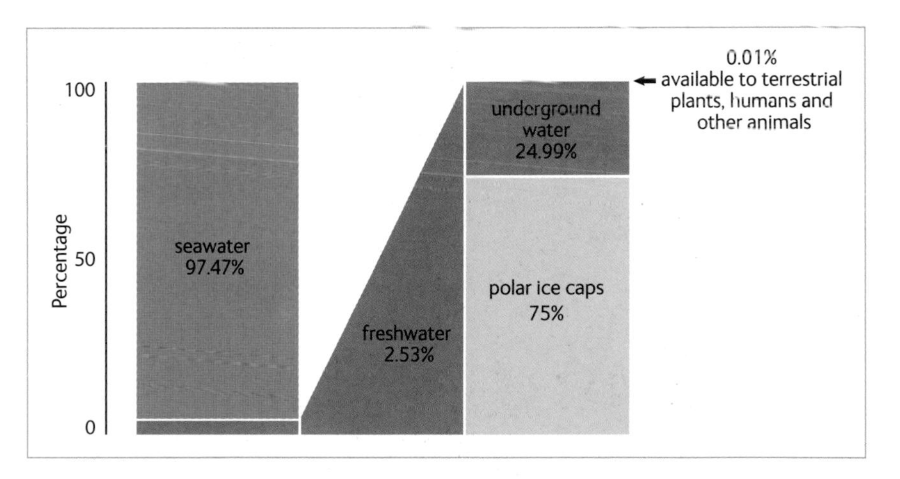

ABOVE Only a tiny fraction of the water on our blue planet is available freshwater.

Humans extract water from natural sources, such as lakes, rivers and groundwater, to provide domestic water supplies, for industry and for agriculture. Much the most

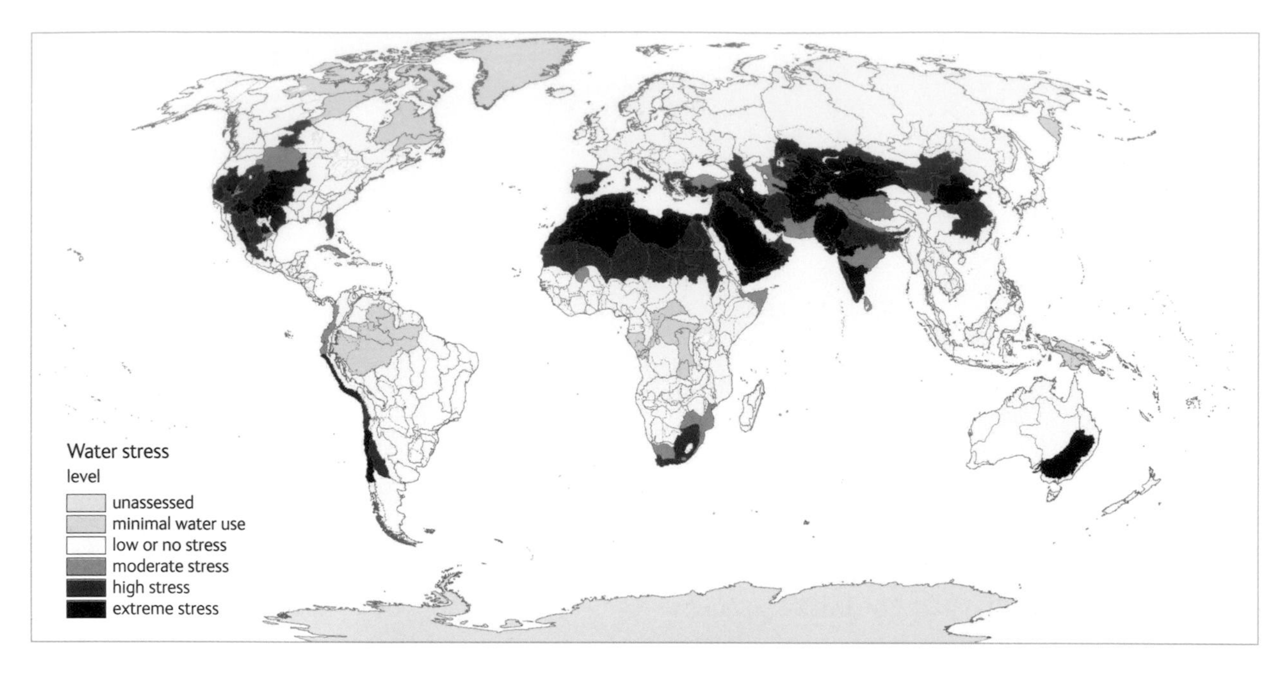

ABOVE Water withdrawal in relation to water availability across the world. The darkest areas are regions of greatest stress, where the ratio of water withdrawal to water availability is greatest.

RIGHT The Aral Sea, in fact a huge freshwater lake, in the former Soviet Union has been drained almost dry by the diversion of water for agriculture from the rivers that feed it. The original perimeter of the Aral Sea is indicated by the outline.

significant of these is agriculture; it is estimated that, by 2025, 60% of the water extracted from natural sources will be for agriculture. Water is used in agriculture in two ways: irrigation for growing crops and the watering of livestock. Irrigation is often very wasteful of water, much of it lost to the air by evaporation before it reaches the crops. Rearing livestock is even more voracious in its use of water. It requires around 1,500 l (400 US gallons) of water to grow one kilogram (2.2 lb) of a cereal crop, but 15,000 l (4,000 US gallons) to produce one kilogram (2.2 lb) of beef, for example.

Humans not only affect natural water supplies by the ways in which they use them but also by how they dispose of them. People living in places where there is no sanitation system for cleaning and decontaminating waste water have little choice but to throw it into a river or into the ground. Each litre of water disposed of in this way is estimated to pollute around 8 l (2.1 US gallons) of freshwater. Pollution of natural water systems by human waste is a major problem worldwide and can only be countered by international investment in sewage-processing systems. Recognition that freshwater biodiversity has become the most important conservation priority worldwide is reflected in the recognition by the United Nations of the International Decade for Action, also known as 'Water for Life', from 2005 to 2015.

Freshwater biodiversity

Although they occupy only a small proportion of the world's surface, freshwater habitats are very rich in biodiversity. Indeed in terms of species richness, measured as the number of species in a given area, freshwater scores slightly higher than terrestrial habitats and much higher than some marine habitats. It is estimated that freshwater habitats contain 6% of all species, and 40% of all fish species. Such figures are only estimates because much freshwater biodiversity across the world is very poorly documented, as reflected by the fact that around 200 new freshwater species are described each year. Our ignorance about what lives in freshwater means that many species are being lost that have never been described and documented.

A major reason why the biodiversity of freshwater fish, molluscs and other organisms is so high is that streams, rivers and lakes are generally isolated from one another, preventing populations of these animals in different water bodies from interbreeding. As a result, there is a high degree of endemism in freshwater habitats. Lake Malawi in southern Africa, for example, is believed to contain between 500 and 1,000 endemic species of fish, although only 315 of these have been described so far.

Another feature of freshwater biodiversity is that the size of a habitat does not reflect the biodiversity that it supports in an obvious way. In the UK, for example, a study of water bodies in a lowland, agricultural landscape found that small ponds contain more species of plants and invertebrates than streams or rivers. Moreover, ponds supported more unique species and more uncommon species than streams or rivers. Such findings are important in the context of conservation; in the UK and elsewhere, larger water bodies, such as rivers and lakes, are largely protected by a variety of regulations, but there is very little legal protection for small ponds. Between 1958 and 1988, about 20% of the UK's small ponds were destroyed; in the county of Huntingdonshire, 99% of ponds disappeared over a 40-year period. Over the same period, the condition of many of the UK's larger rivers, notably the Thames, has greatly improved.

Threats to freshwater biodiversity

There are five major threats to freshwater biodiversity arising from human activities: over-exploitation, pollution, flow modification, habitat degradation and invasion by non-native species. All these local factors affect individual lakes, rivers and ponds. In addition, freshwater habitats are affected by global factors that impact on them at a much larger scale. For example, climate change causes increased water flows where global warming is causing glaciers to melt and decreased water flows where changing rainfall patterns cause drought. Another global factor is nitrogen deposition resulting from agriculture, which, as described below, is affecting ecosystems on a global scale.

In many parts of the world, rapidly expanding human populations are dependent on fish, and sometimes reptiles such as terrapins, caught in inland waters, as a source of protein. The biomass of fish caught in freshwater worldwide increased four-fold between 1950 and 2002. The effects of intensive fishing are complex and poorly understood. Initially, fishing effort is focused on larger fish species that quickly become depleted, leading fishermen to use nets with smaller mesh size. Some fish species are fished to extinction, but others adapt, evolving a life history in which they mature at a smaller size. There may also be a process called 'fishing down' in which local human communities, having wiped out larger, carnivorous species at the top of a food chain, shift their attention to smaller, herbivorous fish lower down. In some inland fisheries, intensive fishing has been sustained over long periods, with a few species becoming more common but many other species disappearing altogether.

An important factor in the pollution of freshwater bodies is that they are affected not only by pollutants dumped directly into them, but also by any pollutants deposited in their catchment area, that area of land that drains into a lake, river or pond. In the UK and many other countries, considerable progress has been made in limiting pollution from sources draining directly into rivers, such as domestic drains and industrial plants, but there remains a major threat from a wide variety of chemicals that are applied to the land, such as herbicides, pesticides and fertilizers. The effects of such chemicals are often complex and very subtle. For example, frog tadpoles exposed to levels of the insecticide dichlorodiphenyltrichloroethane (DDT) that are too low to kill them become hyperactive and therefore more conspicuous to the newts that prey on them.

There is increasing concern about one category of pollutant, the so-called endocrine disruptors or 'gender benders', that include a wide variety of industrial and agricultural chemicals that disrupt animals' hormones and breeding capacity, in almost all cases by feminizing males. During the 1970s and 1980s, for example, biologists found alligators in Florida with reduced penis size and low fertility attributable to high levels of polychlorinated biphenyls (PCBs), DDT and dioxin in the local environment. Feminization of male fish also occurs close to sewage outfalls in some of the UK's rivers. Atrazine is the most widely-used herbicide in the world; 30,000 tons of it is sprayed onto farmland in the USA each year and it is widely used in many other countries. It can be detected at quite high levels in streams and rivers that collect run-off from farmland and has been detected at high levels in rain. In a laboratory study, tadpoles of the African clawed frog, *Xenopus laevis,* were reared in water containing atrazine at concentrations similar to those found in some natural water bodies. The tadpoles grew, developed and metamorphosed into frogs, but many of them were hermaphrodites, meaning that their testes contained both egg-producing and sperm-producing tissues. Many of those that were unequivocally male had a poorly developed larynx, the organ with which males produce mating calls. It is suspected that some endocrine-disrupting pollutants may also harm humans, but as yet there is no clear evidence for adverse effects.

The modification of water flow in rivers occurs throughout the world, and is most extreme where river flow is most variable. Rivers are dammed to provide water during times of drought, and their banks are built up to prevent seasonal flooding. The current capacity of all the world's existing dams is 10,000 km^3 (2400 miles^3), which is equivalent to five times the volume of all the world's rivers. Some of the world's largest rivers, such as the Murray River in Australia and the Rio Grande and Colorado River in North America, now cease to flow in places at certain times of year. Dams block the movements of many fish species, especially those, such as salmon, that move between freshwater and the sea during their lives. Changing the flow of rivers has profound effects on biodiversity, altering the types of plants that can grow in them and, consequently, the types of animals that can inhabit them.

BELOW Lake Powell on the border of Arizona and Utah, USA, is part of the Colorado River. The white 'bath-tub rings' show how much the water level in the river has fallen over recent years.

RIGHT Water lettuce is a seriously invasive water plant in Florida USA. It decreases water flow and affects the local ecology.

Biodiversity in many freshwater habitats has been adversely affected by alien species that have been introduced, deliberately or accidentally, by humans. In water bodies that have already been altered and degraded by people, alien species can become highly invasive and may overwhelm native communities of plants and animals. The water lettuce, *Pistia stratiotes*, is a seriously invasive water plant in Florida, where it blocks access to waterways for boats, decreases water flows and smothers native ecological communities. The Nile perch, *Lates niloticus*, was deliberately introduced into Lake Victoria in the 1950s to provide food for local people; it has thrived there and caused the extinction or near-extinction of several hundred native fish species. Crayfish, introduced from North America, have devastated populations of native crayfish in Europe, primarily because they carry a fungal disease to which they are immune but to which native species are highly susceptible. Invasive diseases of wildlife are increasingly being recognized as a major threat to biodiversity, as discussed in chapters 11 and 12.

Climate change and aquatic ecosystems

At present, the long-term effects of climate change on freshwater habitats and their inhabitants are largely a matter of informed speculation, but they are likely to be severe. Increased average temperatures will cause water bodies to warm up – decreasing the amount of dissolved oxygen that they contain and leading to the extinction of species with high oxygen requirements such as trout and salmon. It is predicted that in western North America, the amount of water currently suitable for these important food and sport fish will decline by about 50%.

Warmer temperatures will melt snow and ice in mountains earlier in the year, causing streams and rivers to rise earlier, with adverse effects for species that breed when water flows are at an optimal level for them. The melting of glaciers and reduction of snow in mountains will make it more likely that water flow in many streams and rivers

ABOVE A fisherman paddling through a mass of dead fish on Lake Rei, Amazonia, Brazil. The fish died as a result of a lack of oxygen during a drought.

slows down and possibly stops altogether in the summer. The ecology of static water bodies, such as lakes, is particularly sensitive to temperature change. Algae thrive in water and warmer temperatures will increase the occurrence of algal blooms, which prevent light and oxygen from reaching the plants and animals beneath the blooms. Wetland habitats will tend to dry out if they are subjected to increased temperatures and decreased rainfall. On the other hand, those close to the sea may become flooded, by sea water, as sea levels rise, with far-reaching effects on their ecology.

The nitrogen cascade

Nitrogen is abundant in the global environment, making up 78% by volume of the Earth's atmosphere. It is essential for life, being a component of the proteins from which plant and animal bodies are made. In the past, levels of nitrogen in the soil were more or less stable, maintained by a nitrogen cycle in which bacteria in the soil and in the roots of certain plants played a crucial role. This balance began to change during the Industrial Revolution, with the burning of fossil fuels that released nitrogen compounds into the atmosphere, and has accelerated more recently with the increasing use of nitrogenous compounds as agricultural fertilizers to boost crop production. The nitrogen cycle has become overwhelmed by the 'nitrogen cascade', in which levels of nitrogen in the soil are rapidly increasing. Nitrogen compounds are washed off the land into streams and rivers where they cause algal blooms. Some blooms contain cyanobacteria, which produce toxins, killing water life and posing a health threat to people. Algal blooms can destroy the fish stocks on which local people depend for food and it can take many years to clear them and restore the ecosystem to normal.

Increased levels of nutrients in water bodies, resulting from increased levels of nitrogenous compounds, favour certain types of organisms, including many that cause disease. There is evidence that increased nutrient levels in water have increased the abundance of a variety of pathogenic (disease-causing) microbes, including those that cause diseases of corals and fish, as well as the human disease cholera. Less directly, increased nutrient levels favour a number of organisms that transmit diseases from one host to another, notably freshwater snails and mosquitoes.

Conserving freshwater biodiversity

Freshwater biodiversity is in crisis. The Living Planet Index that takes into account both the number of species and the abundance of those species in different kinds of habitat (*see* chapter 1) is falling faster for freshwater habitats than for any other part of the global ecosystem, faster even than for coral reefs or tropical forests. Between 1970 and 2000, it fell by 50%.

UNESCO predicts that half of the human population of the Earth will be short of usable freshwater by 2025. It has been predicted that this insufficiency will lead to wars between nations over access to water sources, a scenario that may have been exaggerated. In recent times, nations have generally avoided war over water resources,

RIGHT Water for domestic use has to be carried, often over long distances, in many parts of Africa.

even while fighting over other issues. During three recent armed conflicts between India and Pakistan, for example, their agreement over the use of water from the Indus river continued to be observed.

There will, inevitably, be increasing conflict between the needs of people and the needs of biodiversity for the clean freshwater on which both depend. There will have to be compromises between these conflicting needs in what has been described as a new paradigm in conservation, called *reconciliation ecology*. This involves a scientific approach to accommodating wild species within human-modified or human-occupied landscapes. In southern Africa, for example, already scarce water resources are threatened by an expanding human population, habitat degradation and, in the future, predicted reduced rainfall because of climate change. In the Western Cape of South Africa, the 'Working for Water' project (*see* chapter 7) has been successful. Clearing invasive trees from even modest stretches of river margin has significantly increased the flow of rivers in the Cape as well as restoring the natural 'fynbos' vegetation (*see* chapter 3) and providing much-needed employment for local people. Are there more such win–win solutions to the threats faced by biodiversity all over the globe?

11 CHAPTER Going, going gone: the sixth extinction

EXTINCTION IS A NATURAL PART OF THE PROCESS OF EVOLUTION. It is the fate of all species that they eventually become extinct, and the rich biodiversity that exists on Earth today represents only a tiny fraction of all the species that have ever existed. For example, there are currently about 10,000 extant bird species, but estimates, based on the fossil record, for the total number of birds that have ever existed range from 150,000 to 1,634,000. This disparity arises from the fact that the fossil record is very incomplete. Many species that have become extinct have left one or more descendent species but true extinction is when a species leaves no descendants. The fossil record is rich in large groups of species that became extinct in this sense; they include the ammonites, the trilobites and the dinosaurs.

During the course of evolution on Earth, there have been periods when the rate of extinction was unusually high; there have been five such events, called mass extinctions (see chapter 2). Although these events had a dramatic effect on biodiversity at the time, it is estimated that more than 90% of the extinctions that have occurred during Earth's history occurred outside mass extinction events. Most extinctions are therefore an intrinsic part of the evolutionary process, in which extinct species are replaced by descendent species, a process called 'background extinction'. Since modern humans appeared on Earth, the rate of extinction has increased abruptly and it is now many times greater than the background rate. We are therefore witnessing the sixth mass extinction.

Human impact on planet Earth

The sixth mass extinction event has been underway for a long time, in human terms, but for a short time on the geological time scale. As long ago as 1993, the Harvard biologist E O Wilson estimated that about 30,000 species are going extinct each year, equating to three species per hour. The extinction rate has probably increased since then. Other biologists, however, have produced less alarming estimates. Since 1600, when extinctions began to be recorded, 485 animal and 584 plant species have become extinct, representing an extinction rate of 2.7 species per year, only slightly above the background rate. It is likely, however, that recorded extinctions, which mostly involve well-known and high visibility organisms, such as birds and mammals, do not accurately reflect what is actually happening. There is evidence that extinction rates in groups of organisms that have been less well studied, such as insects and other invertebrates, are much higher than they are in mammals.

OPPOSITE Megafauna mammals from the Pleistocene epoch, 1.8 million to 10,000 years ago, in Australia. The mammals are *Diprotodon* (background), *Palorchestes* (mid-ground) and *Zygomatrus* (foreground).

BELOW Humans have a unique capacity to slaughter other species in huge numbers. This pile of American bison skulls, photographed in 1870, is the result of hunting for food, for sport, and to deprive Native Americans of their primary source of food.

ABOVE Artist's impression of *Procoptodon*, a giant kangaroo found in Australia during the Pleistocene epoch.

The five previous mass extinction events were caused by physical events that created climatic and other changes much greater than the environmental variations to which most animals and plants are typically adapted (*see* chapter 2). The sixth extinction is caused by the direct impact of humans on the ecology of the planet and is thus the first mass extinction to have a biological, rather than a physical cause. Nonetheless, the effects that humans are having on the planet are rather similar to those that occurred during the Cretaceous mass extinction 65 million years ago. In particular, human activities are causing major changes to the climate of the Earth.

The sixth extinction can be divided into two phases. The first began about 100,000 years ago, when the first modern humans emerged from Africa and began to colonize other parts of the world. The second phase began about 10,000 years ago when humans turned from a hunter-gatherer lifestyle to one dependent on agriculture. Able to support themselves by growing crops and domesticating animals, agricultural societies were no longer so dependent on wild plants and animals and began to destroy natural habitats, a process that has accelerated as the human population has grown.

The most spectacular extinctions in the first phase were very large land animals, called megafauna, that inhabited all the world's larger land masses before humans arrived. About 40,000 years ago, humans colonized Australia and rapidly wiped out its megafauna, including a giant wombat (the hippopotamus-sized *Diprotodon*), the flightless thunderbird and giant lizards and snakes. A giant kangaroo standing 2 m (6 ft) tall and weighing over 100 kg (220 lb) survived a few thousand years longer in Tasmania, but then disappeared from there too when humans eventually reached that island refuge. Elsewhere, the same pattern was repeated. The megafauna of North America and South America survived the rigours of the last glaciation but then most of the large mammals went extinct 13,000 years ago just as it ended. This coincided with the arrival of humans from Asia. Madagascar was not colonized until about 2,000 years ago but quickly lost its indigenous elephant birds, a species of hippo and larger species of lemur. The giant flightless birds of New Zealand survived until about 700 years ago when the Polynesian ancestors of the Maoris first arrived.

Vulnerable islands

The destructive impact of human colonization on native species is most evident on small, remote islands, particularly among birds. Ninety-one per cent of the bird species listed as having gone extinct since 1680 were island-living birds. The dodo of Mauritius is a classic example. Birds are able to colonize remote islands that are inaccessible to flightless groups, such as mammals. Over millions of years, many island birds lost the ability to fly, no longer needing it to escape predators. As a result, when humans arrived,

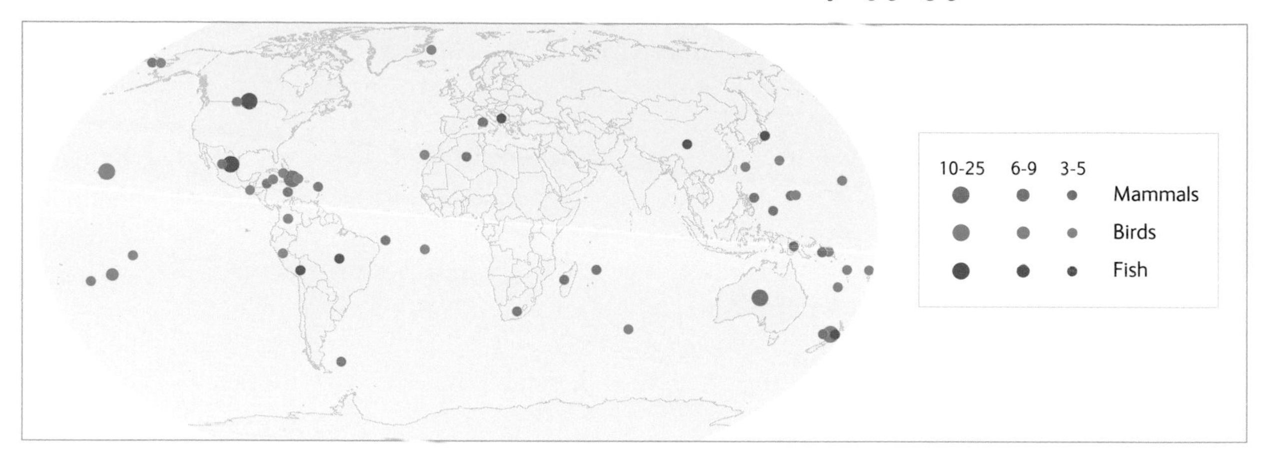

ABOVE The geographical location of species known to have gone extinct since 1600. Many lived on oceanic islands.

they were an easy source of food. Human colonists of Mauritius and other islands have also been very destructive in exploiting the habitat, particularly by destroying woodland.

More importantly, human colonists took with them a number of alien species, such as goats, rats, dogs and cats. Released into a habitat where they had no competitors, such species have had a catastrophic effect on the ecology of many remote islands. Goats destroy native vegetation and rats, dogs and cats prey on largely defenceless

LEFT The extinction of the dodo in Mauritius coincided with the colonization of the island by humans.

native birds. The native birds of New Zealand have suffered many extinctions, largely as the result of two major invasions by rats, the first by Polynesian islanders around a thousand years ago, the second by European colonists in the eighteenth century. Humans and their companion animals have also introduced alien diseases to remote parts of the world. Early colonists of Hawaii brought with them mosquitoes carrying avian malaria, to which native birds had never been exposed. This disease led to the extinction of nearly all of Hawaii's native birds living at altitudes below 600 m (1968 ft). Above this altitude the mosquitoes are unable to live.

Amphibians in crisis

Until recently, most attention has focused on extinction among birds and mammals, groups that have been well-documented. Recently, however, there has been much attention given to the dramatic population declines among amphibians (frogs, toads, newts, salamanders, etc.). It has become apparent that our knowledge of this group of animals was very poor. In 2002 there were 5,399 documented amphibian species. Today, that figure has reached over 6,500 and is still rapidly increasing. A recent global assessment of amphibians concluded that 1,940 (32%) of 6,097 species are threatened by extinction. While scientists first became aware of amphibian population declines in 1989, retrospective analysis reveals that the process was already under way in the 1960s.

BELOW The extinct golden toad, *Bufo periglenes*. Once abundant within Costa Rica's Monteverde Cloud Forest Reserve, it has not been seen since 1986.

ABOVE Mass mortality among frogs in a pool in California, caused by the disease chytridiomycosis.

Amphibians have a complex life cycle, involving two stages that occupy very different ecological niches. They begin life as larvae (tadpoles in frogs and toads) that live at high densities in water and feed on algae and other plant material. Adult amphibians live largely on land, widely dispersed and feeding on insects and other invertebrates. As a result, amphibians face very different environmental threats at different stages of their lives.

The primary cause of amphibian declines and extinctions is habitat loss and degradation. Amphibians typically breed in small ponds, streams and wetlands that are increasingly being destroyed to make way for agriculture and human settlements. Deforestation, particularly in the tropics, is destroying habitat for adults. The thin, permeable skin of amphibians, both as tadpoles and adults, makes them very sensitive to a variety of environmental threats, such as chemical pollution, elevated ultraviolet radiation, diseases and climate change. In many parts of the world, these factors act synergistically. For example, in northwest North America, climate change has reduced water levels in breeding ponds. Consequently amphibians are forced to lay their eggs nearer to the water surface, increasing their exposure to ultraviolet radiation from the sun. This damaging radiation reduces the ability of amphibian embryos to resist fungal diseases.

To this potent mix of environmental threats caused by human activities has been added a natural threat to amphibians, a lethal skin disease called chytridiomycosis. This disease, which may be native to Africa, has become widely distributed throughout the world and has devastated many amphibian species. It affects adult amphibians – tadpoles are not affected by chytridiomycosis but do carry it on their skin. It has been suggested that other factors, such as pollution and climate change, have reduced the ability of amphibians to make an effective immune response to this fungal disease. It is thought that the remarkably rapid spread of chytridiomycosis around the world is the result of human activities, particularly moving amphibians from one part of the world to another.

Many amphibians can only breed successfully in water bodies that are devoid of fish, which eat their eggs and larvae. In many parts of the world, alien fish have been introduced, either to provide food and sport for people, or to control mosquitoes, with a catastrophic effect on amphibian populations. Two amphibian species have been introduced to parts of the world where they have become invasive aliens, with devastating effects on native amphibians. The Central American cane toad was introduced into Australia to control sugar cane pests and has spread throughout the continent, out-competing and often eating native frogs. The North American bullfrog has been introduced into many other continents to provide food in the form of frogs' legs. The bullfrog not only competes with and eats native frogs, but also carries, but is not affected by, chytridiomycosis.

The harmful effects of environmental threats on amphibians are not always apparent as mass mortality among eggs, larvae or adults. A wide range of chemical contaminants, for example, cause sub-lethal effects, such as reduced tadpole growth and severe deformities, including missing legs or extra legs, when tadpoles metamorphose into adults. Such deformities are also caused naturally by tiny parasitic worms, and research

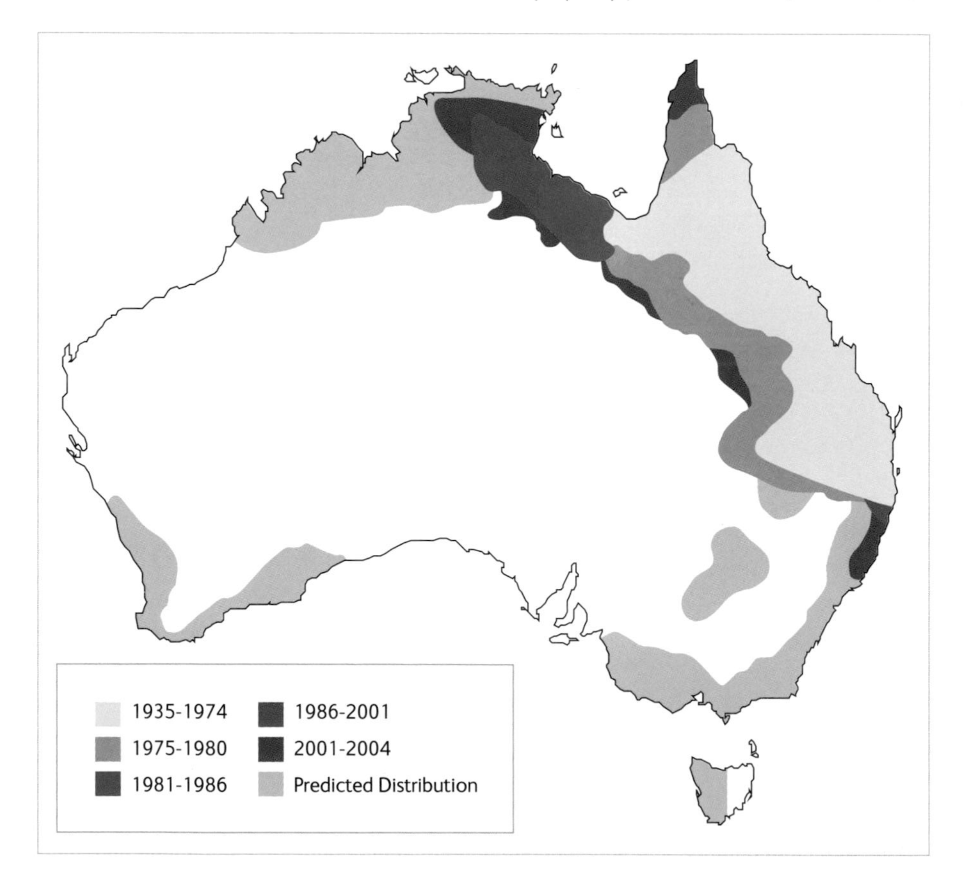

RIGHT Spread of the cane toad, *Chaunus (Bufo) marinus*, across Australia.

LEFT Two normal and four severely deformed newly metamorphosed American wood frogs, *Rana sylvatica*. Two individuals have missing legs; two have extra legs.

has revealed that low levels of man made chemicals reduce the ability of tadpoles to expel these parasites. The widely used herbicide atrazine causes abnormal development in amphibians, causing male frogs to become hermaphrodites, with female tissue in their testes (*see* chapter 10).

Species at risk

It is clear from recent extinctions that certain species are more likely to become extinct than others. Larger species tend to be more vulnerable than smaller species, and those with specialist habits or very specific habitat requirements are at greater risk than those with generalist habits or those that can live in a variety of habitats. In any community of animals and plants, some species are intrinsically rare, and so are obviously at greater risk of extinction than more common species. Abundance does not protect a species against extinction, however. One of the most abundant

RIGHT The passenger pigeon, *Ectopistes migratorius*, of North America used to be one of the most abundant birds on Earth until the arrival of European settlers.

birds ever to inhabit the Earth was the passenger pigeon of North America. When European settlers arrived, they recorded flocks of pigeons so large that they darkened the sky for several days as they passed over. Despite its abundance, the passenger pigeon was driven to extinction, by a combination of hunting and deforestation, the last individual dying in Cincinnati Zoo in 1914. The IUCN maintains Red Lists of species that are threatened with extinction and has defined a five-point scale by which to measure the degree of threat for different species (*see* below). Red-listing a species signals that action needs to be taken to save it. One of the most important ways to do this is through protecting a species' habitat.

IUCN RED LIST CATEGORIES

A long list of criteria is used to place species into five different categories of threat. These criteria include numerical estimates of observed or expected declines in population size, reductions in geographic range, small total population size and fragmentation of populations into small sub-populations. The five categories are:

CRITICALLY ENDANGERED (CR)
A species is Critically Endangered when the best available evidence indicates that it is facing an extremely high risk of extinction in the wild.

ENDANGERED (EN)
A species is Endangered when the best available evidence indicates that it is facing a very high risk of extinction in the wild.

VULNERABLE (VU)
A species is Vulnerable when the best available evidence indicates that it is facing a high risk of extinction in the wild.

NEAR THREATENED (NT)
A species is Near Threatened when it has been evaluated against the criteria but does not qualify for Critically Endangered, Endangered or Vulnerable now, but is close to qualifying for or is likely to qualify for a threatened category in the near future.

LEAST CONCERN (LC)
A species is Least Concern when it has been evaluated against the criteria and does not qualify for Critically Endangered, Endangered, Vulnerable or Near Threatened. Widespread and abundant taxa are included in this category. These species are not on Red Lists.

There are also two extinction categories used in Red Lists:

EXTINCT (EX)
A species is Extinct when there is no reasonable doubt that the last individual has died.

EXTINCT IN THE WILD (EW)
A species is Extinct in the Wild when it is known only to survive in cultivation, in captivity or as a naturalized population (or populations) well outside the past range.

Protecting biodiversity

All species depend upon others for survival, so the most effective way to conserve biodiversity is within ecosystems that preserve intact the whole ecological web. This is easier to do in protected areas such as nature reserves than in places where servicing the needs of humans is the priority, although the two can be reconciled more often than one might think, as for example in the conservation of native 'fynbos' in watersheds in the Cape of South Africa (*see* chapter 10). The Cape is a hotspot of plant diversity, supporting 20% of the plant species found in Africa in just 1% of the continent's area. A surprisingly high proportion of the world's known species occur in hotspots like the Cape because Earth history and evolution (*see* chapter 3) and human erosion of natural habitats have concentrated large fractions of the world's terrestrial biodiversity within a comparatively small area of the land mass. The charity Conservation International uses this fact as a tool to direct its substantial budget towards the protection of the most species for each dollar it spends. Conservation International uses two criteria to define a hotspot: it must contain at least 1,500 species of plants that occur nowhere else (i.e. endemics), and it has to have lost at least 70% of its original habitat.

An evaluation published in 2005 identified 34 biodiversity hotspots that between them contained over half of all known species of plants in only 2.3% of Earth's land surface. The same 34 hotspots also contain over 22,000 terrestrial vertebrate species (amphibians, reptiles, birds and mammals), which is 77% of all those that are known. It is salutary to realize that so much of the world's biodiversity is to be found concentrated within habitats that (by definition) have already been eroded by at least 70%. Not surprisingly, hotspots are home to large proportions of the species on the IUCN's Red Lists. Nearly 60% of the Critically Endangered mammals in the world, 78% of Critically Endangered birds and 85% of Critically Endangered amphibians are endemic to the hotspots. Similar proportions of all Threatened species are also to be found there.

BELOW Map of biodiversity hotspots defined by Conservation International.

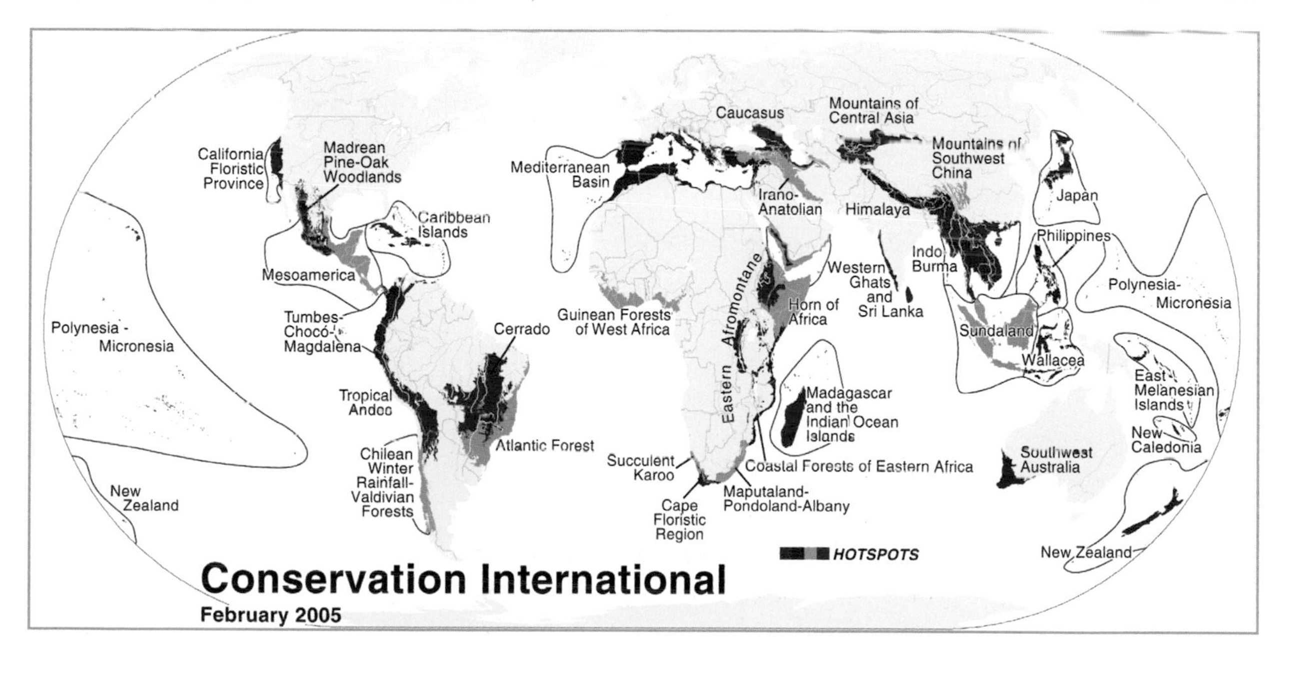

The hotspot concept has helped to focus action on where it is needed most. Through negotiations with governments, often backed up by financial assistance, the area under some form of protection increases year-by-year. For example in 2008, over 250,000 km² (100,000 miles²) of new terrestrial and marine protected area were declared. All biodiversity hotspots are inhabited by people and some, such as Japan, even have high population densities. Conservation actions therefore have to be planned at local level to take account of human needs, particularly in less developed countries where ways have to be found for sustainable development and conservation of biodiversity to go hand-in-hand (*see* below).

ACTION IN BANGLADESH

Bangladesh is one of the poorest countries in the world, yet it is richly supplied with fresh water, fertile soils, fish and wildlife. Some of the poorest of the poor are the communities directly dependent upon fishing, where people compete with one another to sub-let fishing rights from concessionaires who are wealthy enough to buy annual licenses from the government. This system encourages overfishing which, combined with human population growth, pollution and drainage for agricultural development, diminish the supply of freshwater fish. Between 1995 and 2000, the consumption of freshwater fish by the poorest fifth of the population of Bangladesh fell by nearly 40%. In 2000, IUCN classified 40% of freshwater fish in Bangladesh as threatened with extinction. Ten years on, a pilot project in northern Bangladesh has shown how the situation can be completely changed by giving communities 10-year leases on fishing rights in return for adopting conservation measures and sustainable fishing practices. For some villagers, alternative livelihoods to fishing have been financed by micro-loans. Protected areas have been established where fishing is banned and these are patrolled by the community. This protection of a breeding stock has enabled fish catches to increase dramatically and permitted the successful re-introduction of two locally extinct fish species. The pilot project has been so successful in reducing poverty and restoring nature that it is now being introduced to other areas of Bangladesh.

BELOW Sanctuaries help fish yield.

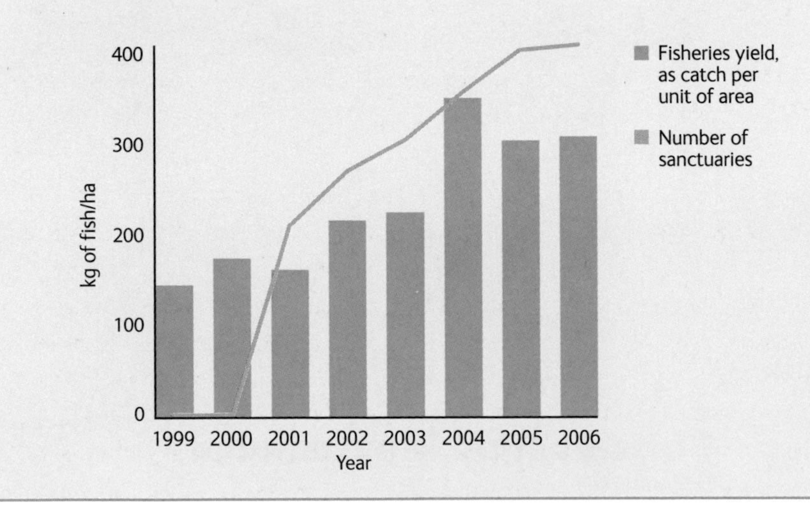

Protected areas do not always live up to the name. An ominous feature of the amphibian decline phenomenon is that many species have declined, and some have become extinct, within such areas. The iconic golden toad, along with several other frog species, became extinct in Costa Rica's Monteverde Cloud Forest Reserve in 1986; it was probably a victim of chytridiomycosis. Yosemite National Park in California has lost much of its once abundant frog fauna, largely as a result of agricultural pollution from farmland a hundred miles away. It is clear that protected areas do not protect wildlife from factors that can cross boundary fences, such as chemical pollution and disease.

Infectious diseases are increasingly being seen as a major threat to rare and endangered species. For example, lions and the endangered African wild hunting dogs, *Lycaon pictus*, in the Serengeti are very susceptible to rabies, which they can catch from managed and feral domestic dogs living just outside the park. An extensive vaccination programme among these dogs seeks to protect wild animals within the Serengeti. Very recently, cancer has been identified as a threat to such wild animals as sea turtles and sea lions. It is likely that such creatures are being infected by viruses and toxins similar to those that cause some forms of cancer in humans.

The convention on biological diversity

The changing state of Earth's biodiversity, and of efforts to protect it, is being monitored under the aegis of the international Convention on Biological Diversity. A target has been set, the year 2010, by when it is hoped there will be a detectable slowing down of the rate at which biodiversity is being lost. At the time of writing this book, the final assessment against the 2010 target has not been released, but the evidence that will go into that assessment is contained in publicly available reports that individual countries submitted in 2009 and these paint a global picture that is at best mixed. As we have seen, the plight of amphibians worldwide is dire. The non-human primates, our closest relatives, are also severely threatened. The National Reports contain some candid assessments of the state of nature. China, for example, talks of the nation facing severe biodiversity loss and degradation of 90% of its grasslands. The United Kingdom's report is more positive about the future, but nature there is emerging from 'rapid declines in biodiversity in the UK during the last quarter of the 20th century'. The general picture is of increasing efforts at conservation in the face of increasing threats to biodiversity from habitat degradation, invasive species and climate change. No country has claimed that the main 2010 target to significantly slow the rate of species extinction will be met. The USA signed, but did not ratify the Convention on Biological Diversity. However, a report published in 2000 concluded that fully one-third of species in the USA are at risk, with 7% being 'critically imperilled'.

The Convention on Biological Diversity provides for National Biodiversity Strategies and Action Plans to be been drawn up by the 166 countries that have signed the Convention. These plans identify how biodiversity will be protected. For example, the plans for the UK name over a thousand species of animals and plants, including 51 species that appear on the IUCN global Red List. Marine species, including fish and mammals, figure prominently on the UK's Red List. The history of fishing makes it all too clear why.

RIGHT Extracts from national reports on biodiversity submitted in 2009 under the Convention on Biological Diversity from a sample of seven countries that have ratified the Convention.

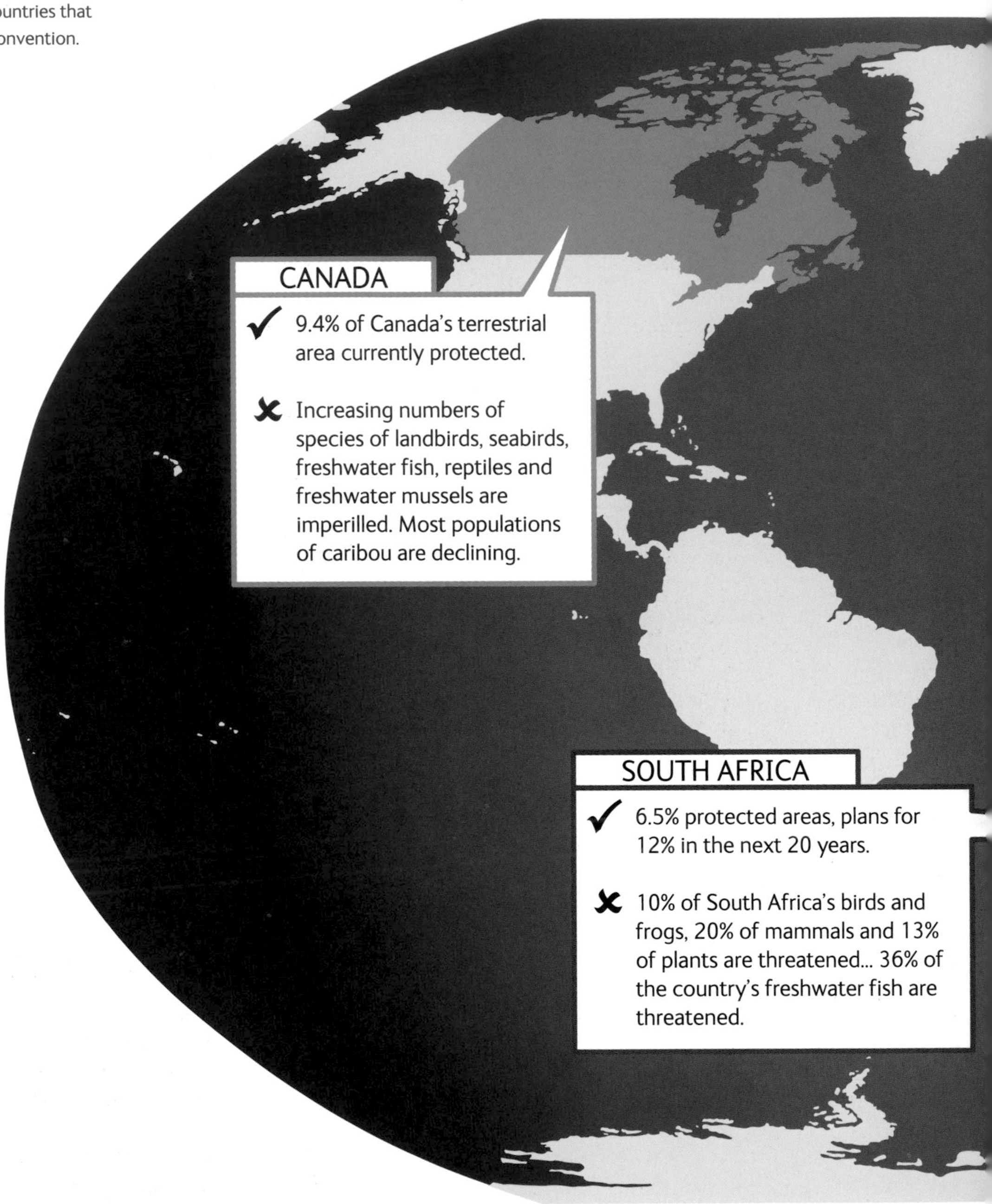

CANADA

✓ 9.4% of Canada's terrestrial area currently protected.

✗ Increasing numbers of species of landbirds, seabirds, freshwater fish, reptiles and freshwater mussels are imperilled. Most populations of caribou are declining.

SOUTH AFRICA

✓ 6.5% protected areas, plans for 12% in the next 20 years.

✗ 10% of South Africa's birds and frogs, 20% of mammals and 13% of plants are threatened... 36% of the country's freshwater fish are threatened.

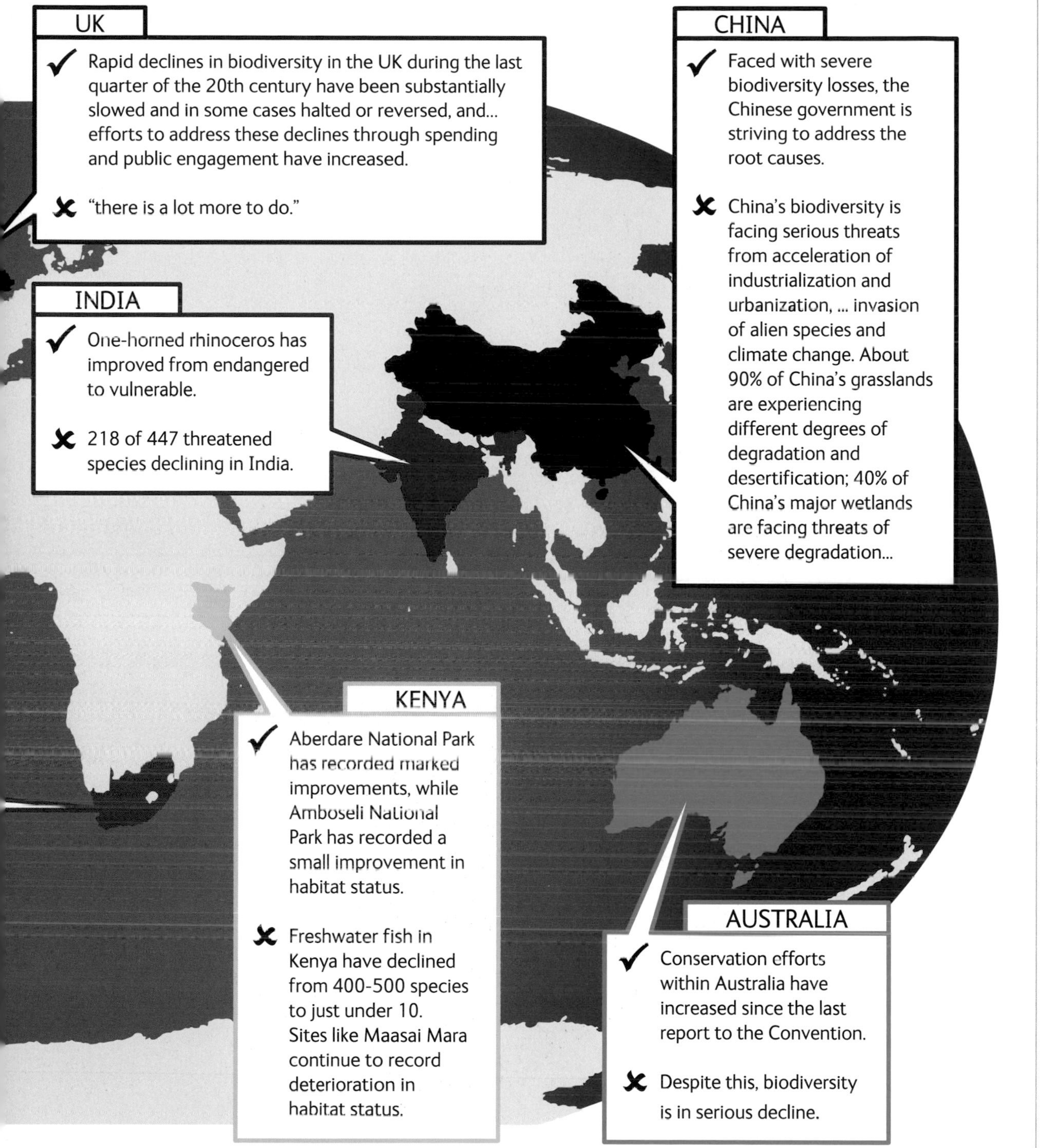

UK

✓ Rapid declines in biodiversity in the UK during the last quarter of the 20th century have been substantially slowed and in some cases halted or reversed, and... efforts to address these declines through spending and public engagement have increased.

✗ "there is a lot more to do."

CHINA

✓ Faced with severe biodiversity losses, the Chinese government is striving to address the root causes.

✗ China's biodiversity is facing serious threats from acceleration of industrialization and urbanization, ... invasion of alien species and climate change. About 90% of China's grasslands are experiencing different degrees of degradation and desertification; 40% of China's major wetlands are facing threats of severe degradation...

INDIA

✓ One-horned rhinoceros has improved from endangered to vulnerable.

✗ 218 of 447 threatened species declining in India.

KENYA

✓ Aberdare National Park has recorded marked improvements, while Amboseli National Park has recorded a small improvement in habitat status.

✗ Freshwater fish in Kenya have declined from 400-500 species to just under 10. Sites like Maasai Mara continue to record deterioration in habitat status.

AUSTRALIA

✓ Conservation efforts within Australia have increased since the last report to the Convention.

✗ Despite this, biodiversity is in serious decline.

DEMARES
DEMARES

12 The exhaustible sea

CHAPTER

OCEANS AND SEAS COVER 71% OF THE PLANET and seem like an inexhaustible resource, but the history of their exploitation shows how their limits have been reached. Commercial sea fishing dates back to Antiquity in the Mediterranean and Black Seas. Ancient Greek jugs, Roman mosaics and Minoan wall paintings reveal an enthusiasm for seafood that spanned shellfish, octopus, lobster and many kinds of fish from mullet to the mighty tunny (better known today as bluefin tuna). Expansion of the Roman Empire brought to the people of northern Europe some Latin tastes in sea fish, such as the pungent fish sauce *garam*, made from rotted anchovies. After the Roman Empire fell, however, archaeological investigations of food remains indicate that northern peoples reverted to catching most of their fish from freshwater.

BELOW Ancient Greek vase showing the processing of a bluefin tuna, *Thunnus thynnus*.

Just after the end of the first millennium a sea-fishing revolution swept northern Europe, triggered it seems by the world's first fisheries crisis. A rising human population and the spread of Christianity increased demand for fish (because Christians prohibit the consumption of meat from quadrupeds on certain days – 150 or more days for the most devout). At the same time, rivers suffered increased sediment pollution from land clearance for agriculture, and a proliferation of mill dams blocked the upriver migration routes of key fishery species such as sturgeon, salmon and whitefish. Together these habitat changes led to a collapse in freshwater fish supplies that prompted people to set forth in boats to catch fish at sea. It also led to the development of freshwater aquaculture in northern Europe, particularly for carp.

Sea fisheries thereafter prospered for hundreds of years. As time went on, people adapted to changes in marine fish demand and supply with new inventions such as bottom trawling in the fourteenth century and long-line fishing in the eighteenth century. Old methods of hand gathering and intertidal traps, in use since pre-history, continued alongside these advances and provided good returns well into the late nineteenth century.

OPPOSITE Sea fisheries have prospered for centuries until now. A rising human population and an increased demand for fish have seen the world's oceans pushed to their limits.

Roads, railways and cheap fish

Fishers are often caricatured as pessimists because of their perennial complaints about falling numbers and sizes of fish. Over the long period of fisheries prosperity between the eleventh and eighteenth centuries, there were occasional outbreaks of woe in European fisheries that have left their mark in legal or parliamentary records. However, the grumbling really took off in the first half of the nineteenth century when the construction of swift road and rail connections between coastal ports and inland cities made fresh fish available to a larger population and demand increased.

In the UK, fishers were especially vexed by the spread of bottom trawls pulled across the seabed by sailing boats. Those using hook, net or trap to catch fish argued that trawls were wasteful and destructive because they caught huge numbers of undersized fish and tore up seabed habitats formed by delicate invertebrates like corals, sponges and sea-fans. J C Bellamy summed up feelings when he wrote in 1843,

BELOW A mid-nineteenth century sailing smack towing a beam trawl. This simple fishing method dates back to at least the fourteenth century and remained virtually unchanged until the late nineteenth century when trawlers gained steam power.

> *The trawl is still suffered to be employed in all its baneful tendencies without restraint or limitation...Dragged along with force over considerable areas of marine bottom, it tears away promiscuously, hosts of the inferior beings there resident...An interference...of such magnitude, and of such long duration, will hereafter bring its fruits in a perceptible diminution of those articles of consumption for which we at present seem to have so great necessity.*

Despite the criticisms, trawling prospered because there was good money to be made by wealthy investors in positions of influence.

In the late nineteenth century, fishing effort increased dramatically, this time with the addition of steam power to fishing vessels. Steam released fishers from the bonds of wind and tide, increasing the time available for fishing and the size of the boats and nets they could use. The greater speed of steam boats meant that they could fish farther afield and still land fresh fish. By 1920, the use of steam trawlers and drifters had spread to the Arctic, North America, Africa, Asia and Australia. With hindsight, it is clear that the use of motorized vessels came at a cost. Returns from near coastal waters declined following the onset of steam power, forcing vessels to travel ever further in search of good catches. Fishers also adapted to falling supplies by switching to species that had in the past been thrown away because there was no market for them. So, for example, in Europe and North America, dogfish became target species as cod, haddock and other traditional species declined.

LEFT Intact *Lophelia pertusa* coral reef with a redfish contrasts sharply with a similar reef after bottom trawling, below.

BELOW Damage done to deep-water corals by the passage of bottom trawls on deep-water *Lophelia* coral reefs in eastern Canada.

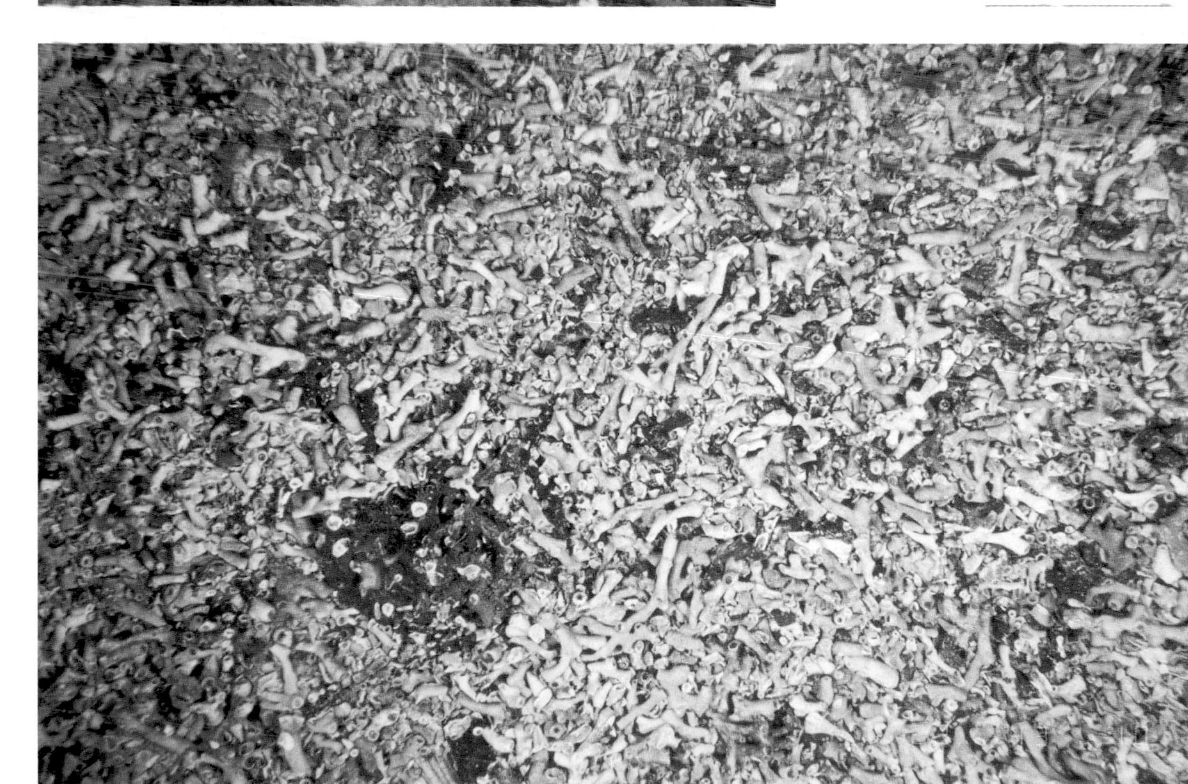

Human ingenuity also helped to prop-up landings as stocks declined. Throughout the twentieth century, fishing power was bolstered by countless innovations, including the adoption of diesel engines in the 1920s, the introduction of monofilament nets and lines in the 1950s, the widespread use of echo-sounders by the 1970s, and the development of a panoply of high technology electronics by the turn of the twenty-first century. In spite of this technical fish-catching prowess, or rather because of it, populations of fish and shellfish went into steep decline worldwide. In the last two centuries, the expansion of fishing power has resisted our capacity or will to restrain it in all but a few parts of the world.

ABOVE Catch from a Grimsby trawler around the end of the nineteenth century. Landings like this from the new steam powered trawlers were once commonplace, but overfishing soon pushed stocks into decline.

Today, our ability to sustain the size of catches by increasing fishing effort and through the use of more sophisticated technology is nearing its end. We now fish virtually everywhere and at depths of up to 3 km (about 2 miles). We exploit nearly all of the desirable and productive species and there are few 'underexploited' species left to switch to. After correcting for systematic over-reporting of catches by Chinese officials, total world fish landings have fallen since the late 1980s, ending a period of continuous increase that lasted for at least 500 years.

Today the news is full of stories about problems affecting fishing. However, when journalists report on the latest story they almost always talk about a stock collapse of this or that species being imminent, but rarely suggest that it has already occurred. In reality, many of the major fish stocks we exploit collapsed long ago. Since 1950, the Food and Agriculture Organization of the United Nations has monitored landings of key fish stocks worldwide. Their records show that two-thirds of all the species we have exploited since then have fallen below 10% of their peak landings (i.e. collapsed).

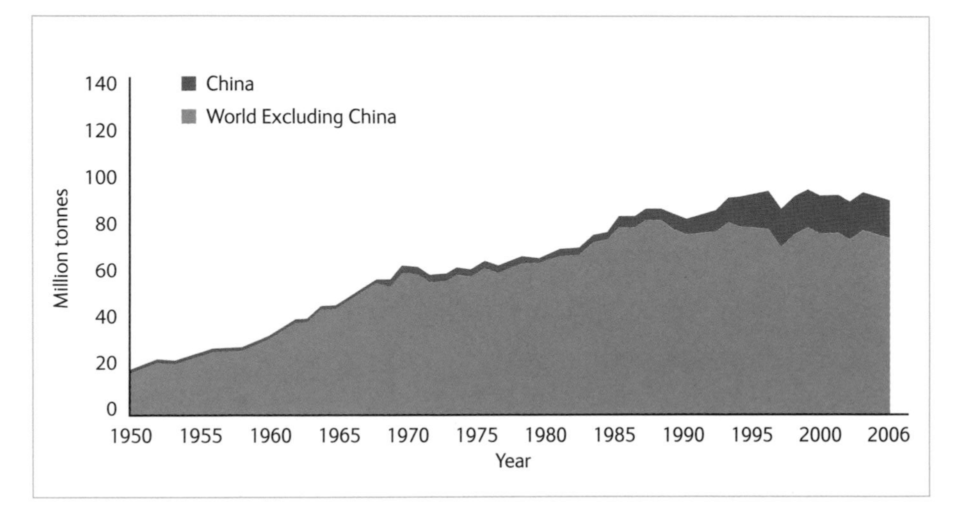

RIGHT Graph of the trend in global landings of wild fish and shellfish over the past half century. When figures are corrected for over-reporting by Chinese officials (lower line), it is clear that catches peaked in 1988 and have been in decline since.

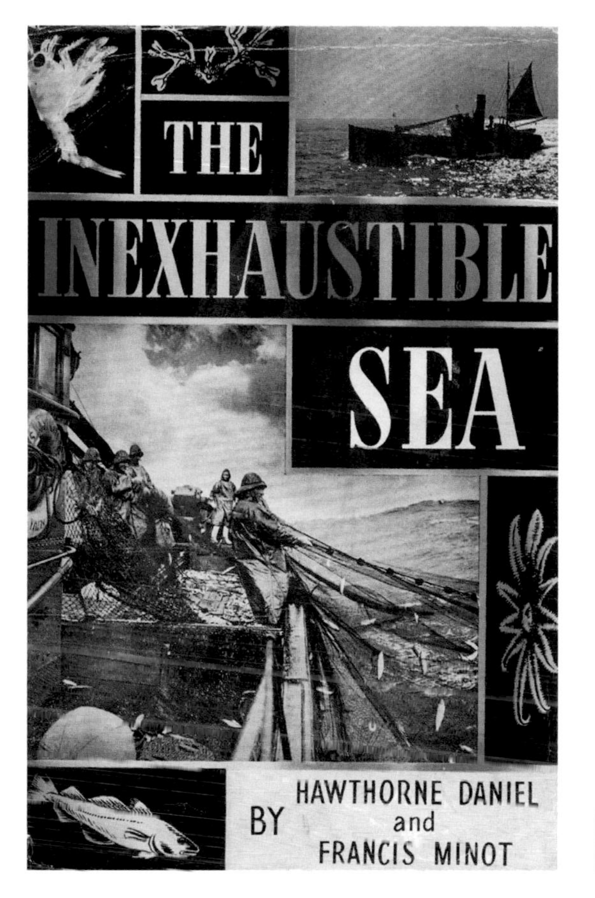

ABOVE Attitudes towards the oceans have changed in the last 60 years. In the 1950s people saw the sea as an inexhaustible source of food for the burgeoning human population. Today, the view is more pessimistic.

Extrapolating this trend forward in time affords a chilling view of a future without fish. If fisheries continue on their present course, all major fish stocks are predicted to experience collapse by the middle of the twenty-first century.

In some circles this research has been criticized for being too simplistic. Certainly, some fisheries have recovered from collapse, and others are on the mend. But as things currently stand such examples are the exceptions not the rule and the logic of the 'doom and gloom' prediction is inescapable. If you take more fish than are produced each year, as we do in most fisheries, then eventually you will run out of fish.

The sea before fishing

Fish are hard to see and harder still to count, which provides one reason why fishery problems have been overlooked or ignored for so long. Scientists measure fish population sizes mainly by a combination of experimental fishing, through records of catch rates from commercial landings, and by tagging studies. Sadly, quantitative estimates of population size are available for a relatively small number of species compared with the variety of animals we exploit and in most cases records go back only 20 or 30 years. Given the long history of fishing, these recent figures give only a partial picture of the extent of the declines suffered. When modern records are supplemented with other kinds

RIGHT Change in fish sizes and composition between 1957 (top) and 2007 (bottom) from recreational catches around Key West in Florida, USA.

of information, such as accounts in historical archives, old photographs, oral history and analyses of fish or shellfish remains in marine sediments, it is possible to gain a much better understanding of the scale of impact that fishing has had on marine life.

Both on land and in the sea, hunters and fishers tend to pursue protein in large packages first. Bigger species tend to be the most desirable and the first to be exploited. Fishing methods are mainly size-selective, and the largest fish tend to be predators, which are bold and quick to take a hook or enter a trap. Larger fish are also retained by nets that let smaller animals escape. Unfortunately, big species tend to have a suite of characteristics that make them highly vulnerable to depletion. They often live long lives, grow slowly and mature late at large sizes. Their late start to reproduction provides many opportunities for capture before they have had the chance to multiply. The result is that intensively fished places lose large species first, a pattern that has been seen in region after region throughout the world.

What were the seas like before fishing? In most places you have to go back to times long before living memory, to understand this because they have been fished for such a long time. Historic accounts offer valuable insights into the past abundance of marine life. Take the 'picked dogfish', for example, a voracious species of shark that can reach 1.5 m (5 ft) long. Jonathan Couch, an English ichthyologist, wrote in 1862,

> *The picked dog is the smallest, but by far the most abundant of British sharks. It is found at all seasons on every part of the coast of the United Kingdom, but in the greatest numbers in the west and south; where at times they exist in such multitudes, as to occupy the full extent of sea for scores of miles; and twenty thousand have been taken in a sean [seine] at one time, without any apparent lessening of the numbers.*

Today, this species has declined to such an extent that it is listed by the World Conservation Union as Critically Endangered in the northeast Atlantic.

BELOW The picked dogfish, *Squalus acanthias*, is a species that was considered a pest in the early nineteenth century because of its habit of taking other fish from nets and hooks. It became a target species in Europe in the early twentieth century when fish and chip shops provided a market for 'nameless' slabs of fish flesh from species that were previously shunned.

Picked dogfish rubbed shoulders in the southwest of England with large numbers of blue, mako and porbeagle sharks, porpoises and dolphins, hake and giant skates. These predators were there because of the breathtaking abundance of prey fish, namely the pilchard. John Murray wrote of pilchards in his *Handbook for Travellers in Devon and Cornwall* in 1851,

> *Pursued by predaceous hordes of dogfish, hake and cod, and greedy flocks of seabirds, they advance towards the land in such amazing numbers as actually to impede the passage of vessels and to discolour the sea as far as the eye can reach... Of a sudden they will vanish from view and then again approach the coast in such compact order and overwhelming force that numbers will be pushed ashore by the moving hosts in the rear... In 1836 a shoal extended in a compact body from Fowey to the Land's End, a distance of at least 100 miles if we take into consideration the windings of the shore.*

Similar pictures of far greater historic abundances of predators and prey can be drawn for almost any region of the world where records exist. Sharks swarmed around the boats of seventeenth century explorers and pirates in the eastern Pacific. Whales were so abundant in Patagonia in the early eighteenth century that they threw water on deck as they blew. Black sea bass weighing hundreds of kilos formed enormous spawning aggregations off the California coast up to the early twentieth century. Alewife, shad and salmon migrating up North American rivers in the sixteenth and seventeenth centuries were sometimes so abundant that there seemed to be more fish than water.

Collapse of the megafauna

Taken together, many lines of evidence indicate that since the onset of fishing, large-bodied animals have declined by 75–90%. Numerous species have seen even steeper collapses, such as the common skate in UK waters, which is now at least 1,000 times less abundant than it was in the nineteenth century, and has been eliminated from large

BELOW Collapse in landings over the twentieth century of several species of megafauna found around the coasts of the UK due to intensive fishing. Landings of skates and rays include at least 10 species. Grouping them together masks much steeper declines by large bodied species like the common and white skates. The UK once had abundant marine megafauna, but overall large animals have been reduced to a few percent of former numbers.

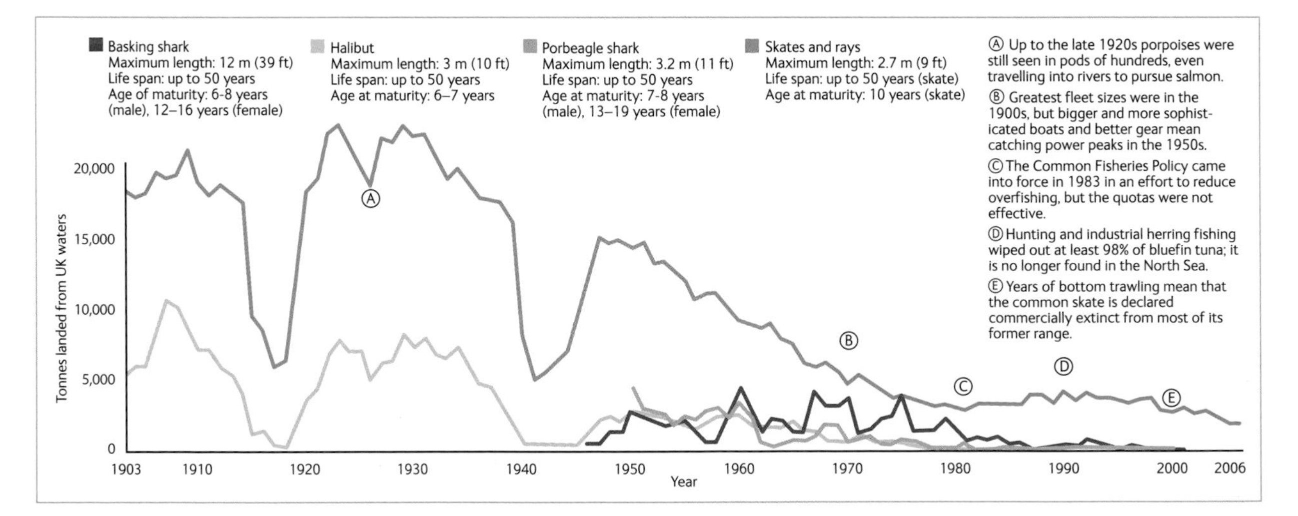

expanses of its former range. The northern cod is over a hundred times less abundant in eastern Canada than it was when Newfoundland was first sighted by John Cabot in 1496. Bluefin tuna in the Mediterranean is down by more than 95% since the 1960s.

Few marine species have yet been driven to complete extinction, but many are on a fast track to oblivion. Losses have been driven not only by direct removals from fishing, but from the use of destructive fishing gears that inflict huge collateral damage on habitats and non-target species. Bottom trawls are considered among the worst gears in use today for exactly the reasons given in the nineteenth century. However, until recently, few people appreciated how bad trawling was in this respect, because most of the damage was inflicted long ago. When trawls first came into widespread use, they transformed vast areas of seabed from places crusted with diverse and three-dimensionally complex communities of invertebrates, to the shifting sand and gravel barrens seen today. People of today's generations only woke up to the damage when trawlers moved into deep water in search of new fish stocks, and encountered once again habitats that had never experienced the pass of a trawl. Nets laden with strange corals and unfamiliar sponges rang alarm bells among conservationists, especially when radiometric aging revealed that many of these animals were hundreds, or even thousands of years old.

Sleepwalking to disaster

There appear to be several reasons why we have let things get so bad in the oceans. The most obvious is that the sea is so enormous that it seemed impossible that humanity could ever deplete its resources. The oceans make up 95% of the volume of living space on the planet so this view seemed hard to fault. However, now that the machinery of modern fishing has grown so far in scope our impacts on the sea are omnipresent.

Another reason is that people are not very good at noticing creeping changes that span human generations. Each new generation grows up in a world that is different from that experienced by the one before. It is human nature to set greater store in personal observations than in the views of our elders, so our appreciation of the state of the sea shifts from generation to generation. Past losses are overlooked and we come to view the environment as less altered or damaged than it has been in reality. The term 'shifting baselines' has been coined to describe intergenerational changes in the way we see our environment. Shifting baselines cause problems when people fail to spot changes that threaten their well-being, such as erosion that reduces soil fertility, or falling water tables, or mangrove loss that destablizes tropical coasts...or fish that are about to run out!

A third reason for things getting so bad in the sea is the tragedy of the commons. With a few exceptions, people do not own areas of the sea or the animals that live in it in the way they own property on land. Where resources are common property, shared by many, there is a tendency for individuals to seek to maximize personal gain by taking as much as they can. Such selfish behaviour multiplied across many people leads to resource

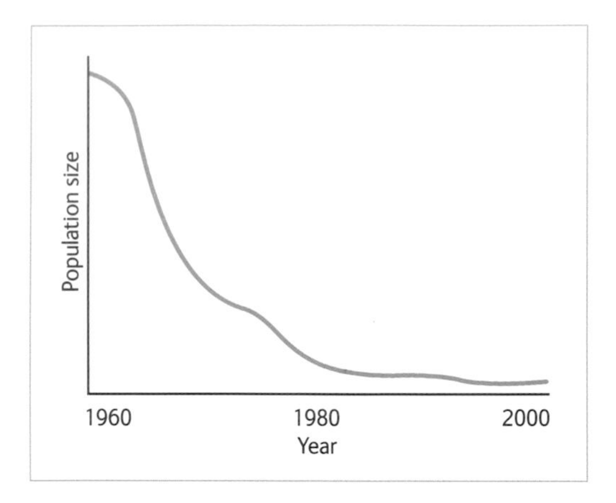

RIGHT Collapse of stocks of bluefin tuna in the Mediterranean. The International Commission for the Conservation of Atlantic Tunas was created in 1969 and has presided over the collapse since then. Its appalling record of incompetence and institutional intransigence has led it to be dubbed The International Conspiracy to Catch all Tunas.

depletion. Collective restraint that ensures sustainable use would leave everybody better off, but is something that rarely happens without good management.

Finally, throughout much of the world, because marine resources are generally common property, the management of them has been vested in politicians, which is a role that does not sit well with some of their other obligations. Political decisions are dominated by short-term pressures, like industry's demands for more fish, and their own desires for votes. In Europe, under the Common Fisheries Policy, fisheries ministers have for the last 20 years set total allowable catches at levels 25–35% above those recommended as sustainable by scientists. If a farmer took a few more sheep to market every year than lambs were born, he would soon have no flock. So it is also with fish. Fisheries decision-making in Europe as currently practiced guarantees stock collapse. The only uncertainty is how long it will take.

Restoring life to the sea

Despite all of the above, not all is lost, because given sufficient protection, marine animals and plants can bounce back. In many countries of the world, marine protected areas have been established where fishing is prohibited. Time and again exploited species have rebounded within such areas, often reaching three, five or ten times greater abundance within a decade of protection, sometimes more. For example, around the Caribbean Island of St Lucia, commercially important fish species increased five-fold inside marine reserves within 7 years of protection. In Kenya, the amount of parrotfish went up 20 times in four marine reserves within 10 years of protection. On Belize's Glover's reef, there were over 20 times more lobsters inside a marine reserve than in nearby fish areas after 10 years of protection.

Protection also works in temperate areas and at the scale of industrial fisheries. In 1995, four large fishery areas were protected from bottom trawling and dredging for scallops on Georges Bank off Cape Cod in the USA. Within a decade, scallops and haddock staged a comeback. Their return highlights an important role for marine reserves in recovering ocean fisheries. If you do not kill animals in the sea, they become more numerous, live longer, grow larger, and produce more young. For most species, eggs and larvae drift on ocean currents, meaning that fish in reserves can re-seed surrounding fishing grounds. Fishery closures off Cape Cod have revitalized the moribund haddock and scallop fisheries. Fishers around St Lucia reserves catch nearly twice as much as they did before protection, and fishers in Kenya set their traps close to reserves to intercept fish spilling out of the protected areas. A further benefit of protection is that habitats can begin to recover from the destructive effects of gears like trawls. To have a real impact on ocean recovery, reserves must be numerous and established in well-connected networks. Best estimates suggest that coverage should be in the order of 30% of the sea or so, to maximize benefits to biodiversity and fisheries alike.

Reserves cannot reverse all of the harm that fishing has done, however. If fish and fisheries are to prosper in future, they must be complemented by other measures. These include reduced fishing pressure in fishing grounds, the restriction or prohibition of the most damaging fishing methods, use of the best available technologies to reduce the collateral impacts of fishing on other wildlife, and the replacement of risk prone decision-making by politicians with precautionary targets based on sound science, made by people who are independent of political or fishing-industry ties. These solutions to overfishing are widely known and are not complicated. We could start to make them today – we just need to find the political will to do it.

LEFT Fishermen in St. Lucia saw a near doubling in catch rates 5 years after the establishment of a network of marine reserves that were protected from fishing. Inside reserves, commercially important fish stocks increased five-fold, while adjacent fishing grounds saw increases of three-fold. Marine reserves help replenish stocks in fishing grounds by emigration of fish and shellfish, and the export of their eggs and larvae on ocean currents..

WALDEN

or,

Life in the Woods

I do not propose to write an ode to dejection, but to brag as lustily as chanticleer in the morning, standing on his roost, if only to wake my neighbors up. —Page 389.

13 CHAPTER Climate change

IN *WALDEN; OR, LIFE IN THE WOODS*, Henry David Thoreau's lyrical account of the year 1845–6 that he spent living in a cabin in the woods at Concord in Massachusetts, he reflected upon life and nature in its smallest details. Towards the end of the book Thoreau wrote 'I am on the alert for the first signs of spring, to hear the chance note of some arriving bird or the striped squirrel's chirp, for his supplies must be now nearly exhausted.' Everything in nature was of interest to Thoreau and no episode of the changing scene was too inconsequential for him to record. 'On the 13th March, after I had heard the bluebird, song-sparrow, and red-wing, the ice [on Walden pond] was still nearly a foot thick.' His private journals and notes contain the dates of first flowering for over 500 species of plants made between 1852 and 1858.

OPPOSITE The title page from Thoreau's book detailing the life and nature surrounding him during his year spent living in the forest.

Meanwhile, in the tiny village of Stratton-Strawless in the county of Norfolk, UK, the Marsham family were into the second century of a tradition begun by their forebear Robert Marsham in 1736. Each spring, members of the family recorded the date when the first snowdrop, the first wood anemone, the first hawthorn and the first turnip bloomed; the date that leaves first appeared on a dozen species of tree; the arrival of the first swallow, cuckoo, nightingale and nightjar and the first frog or toad to issue an amorous croak. This Marsham family tradition was continued for five generations, ending only in 1947. Like the Marshams, Henry David Thoreau and Gilbert White, author of the *Natural History of Selborne*, many educated country dwellers of the eighteenth and nineteenth centuries painstakingly recorded the vagaries of nature's calendar. Although such an interest may seem more rustic than rational to a twenty-first century urbanite, these records from earlier centuries have proved to be of great scientific value. They show that most species are very sensitive to seasonal temperature and comparisons with recent data indicate a strong response to global warming of the climate. In fact, sustained changes in the timing of events in the living world gave the first clear and unequivocal alert that global warming was a reality and not just a blip on the meteorological charts.

ABOVE Robert Marsham and his family recorded the first appearance of plants and animals in their area.

ABOVE AND RIGHT The first signs of spring in Norfolk, UK. A barn swallow, *Hirundo rustica*, (above left) returning to England, European common frogs, *Rana temporaria*, (above right) on their way to the pond to lay eggs and wood anemones, *Anemone nemorosa*, (right) in bloom.

Plants in Concord now flower on average a week earlier than they did in Thoreau's day. A few species like highbush blueberry or yellow wood sorrel flower 3 weeks to a month earlier than they did 150 years ago. Twentieth century records of first flowering dates in England show something even more alarming – the effect of the warming climate is getting stronger. A significant number of wild plants advanced their flowering dates by 15 days in just the single decade of the 1990s. Not only is spring getting earlier, but autumn is falling later, which has lengthened the season during which fungi can be found fruiting. Not necessarily a bad thing if you like collecting them! Indeed, one might ask, does it matter that the growing season is getting longer as the climate warms? Could the changes even be good news rather than bad for biodiversity?

Quick march

It is probably a universal rule that changes of any kind produce winners and losers. The history of life on our planet certainly bears this out (*see* chapter 2), and even among invasive species which in general will be winners, it is expected that some plants, such as cheatgrass in the south of its US range, could retreat in some regions, which would be a good thing for the native flora and fauna. As you might expect, not all species respond in the same way to climatic change and there are already clear signs of which species may be the winners and which will be the losers. If you go hiking in the Green Mountains of Vermont, at about 800 m (2667 ft) elevation there is a rapid shift in the composition of the forest from the trees characteristic of the northern hardwoods (sugar maple, American beech, yellow birch) to a forest dominated by conifers (red spruce, balsam fir) and paper birch. This shift from one forest type to another as you ascend the mountain reflects the lower average temperatures experienced by species at higher elevations. The same shift from northern hardwoods to boreal forest occurs as you travel northwards, also reflecting a cooler climate at high latitudes.

Over the last 40 years of the twentieth century the mean annual temperature in Vermont increased by 1.1°C (2°F) and in the same period, northern hardwood trees advanced upslope while the boreal forest species retreated by about 100 m (330 ft) elevation. In recent decades, plants in Western Europe have been creeping uphill at a similar rate. Such a rapid shift is remarkable for trees that typically live for from one up to several centuries. Young trees can usually only replace older ones by growing in the light gaps created in the canopy when an old tree dies. What might otherwise have

BELOW Changes in forest type that are predicted to occur in the eastern United States as a consequence of climate change.

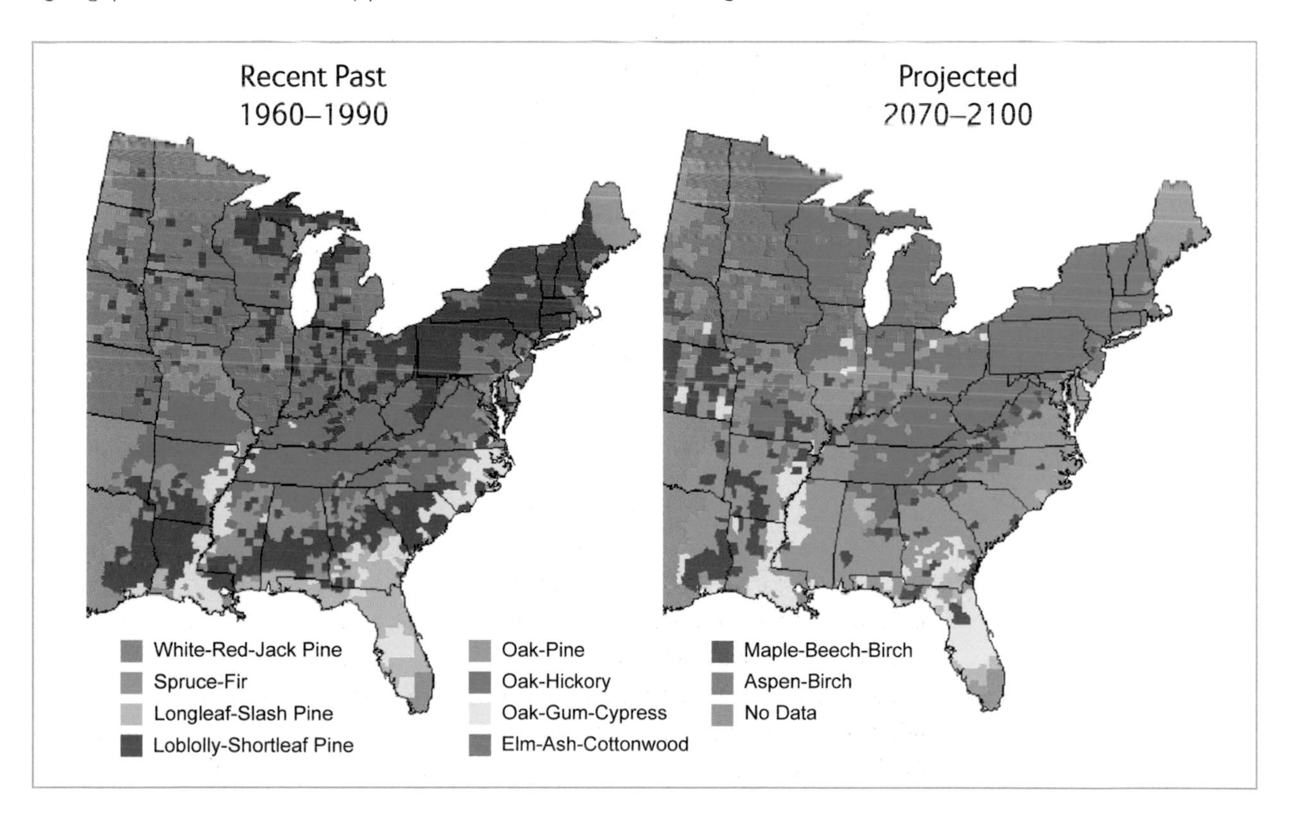

been a slow advance of the hardwood species inhibited by the longevity of established trees was probably accelerated by an increase in tree death rates. Further warming later this century may see northern beech–maple forest itself replaced by other tree species. American beech is in any case severely threatened by beech bark disease that attacks trees infested with a species of non-native scale insect.

An increase in the mortality rate of forest trees has been observed in many other North American forests and this seems to be at least partly linked to climate change. Tree mortality rates and the rate at which trees are naturally replaced in tropical forests seem to have been accelerating too, just as in North America. Exactly why this higher turnover is happening is not clear yet, but the effect is very widespread in the tropics and therefore global processes such as increased atmospheric CO_2 and/or changes in climate are almost certainly involved.

Animals as well as plants are on a forced march to keep up with shifting climatic conditions. In the UK, for example, butterflies, mammals and birds have recently been moving northwards at more than a kilometre a year and dragonflies at twice this rate. In the region of Mount Kinabalu in tropical Borneo, mean annual temperature increased by about 0.7°C in the last four decades of the twentieth century. Over the same period the distributional ranges of moths on the mountain moved upslope by an average of nearly 70 m (2330 ft). It has been predicted that climate change ought to impact upon nature even more strongly outside the tropics because temperature increases are expected to be proportionally greater nearer the poles. Consistent with this prediction, average temperature rose by 1.3°C (2.3°F) in just 30 years in southern Spain and the lower limit of the distribution of certain butterfly species moved 212 m (695 ft) upslope on mountains there.

A northern species that can adapt to global warming by moving to cooler climes is less likely to be driven extinct than a species that cannot adapt. However, there may be obstacles of several types to overcome, even for species that can move. On mountains, species that are driven upwards by warming at the lower edge of their distributional range may find that there is not as much suitable habitat available higher up. The Spanish butterflies lost an average of 30% of their habitat because the higher, cooler slopes were also smaller in area. Losses as great as 80% of these butterflies' habitat could occur over the coming century. A contraction of range has also been observed among the fauna of small mammals living in Yosemite National Park where species have on average moved 500 m (1670 ft) upslope during nearly a century of observation, while temperature increased by about 3°C (5.4°F).

A study of a very wide range of animals and plants in several parts of the world found that range contractions caused by global warming could threaten 11% of species with extinction, even if species were able to move with the climate which warmed only according to the most optimistic forecast. If global warming is more severe, a third of species could be lost. Potential winners may have a problem too because although species of warm conditions ought to have more suitable habitat as the climate warms up, the new habitat may occur in fragments that are difficult to reach by the normal means of dispersal. Many butterflies reach the northern edge of their geographic range in the UK and ought to become more common with global warming as the limits of favourable climate move northwards. In reality, however, farmland, roads and other areas of unsuitable habitat present impassable barriers to many rarer species. A solution to this problem that has proved successful is to deliberately introduce butterflies to uncolonized areas of new habitat. This kind of conservation action can be controversial. How far can species be moved before they should be considered as non-native? Should non-natives be introduced?

Assisted migration

Assisting species to colonize new areas of habitat created by global climate change is already taking place in British Columbia, Canada. Projections of climate change in the province show which areas will be suitable for various forest types in the future, but the climate is changing so fast that trees cannot be expected to colonize quickly enough to keep up. Habitat suitable for ponderosa pine, a southern species, is predicted to advance northwards by 100 km (63 miles) a decade.

For northern species, the speed of change could result in forests of dying trees at the southern edge of the range, with no compensating replacement of populations taking place at their northern edge. In other words, species could be chased by an unsuitable climate that they cannot escape from quickly enough by dispersal. Populations of forest trees tend to be closely adapted to their local climate, so even well within a species' range, it may be necessary to move seeds northwards to enable tree species to survive a changed climate. Fortunately, foresters set up experiments many decades ago to investigate how moving seeds throughout a tree's range affects growth and survival. These studies have shown that the best growth of young trees is often seen in those grown from seeds sourced well to the south of their planting place. However, though

BELOW How ecological zones in British Columbia are predicted to shift as a consequence of climate change. The main types are: green – coastal western hemlock forest, yellow – interior Douglas fir forest, red – bunchgrass grassland, light orange – Ponderosa pine forest, blue – boreal white and black spruce forest, rose pink – Engelmann spruce and subalpine fir forest.

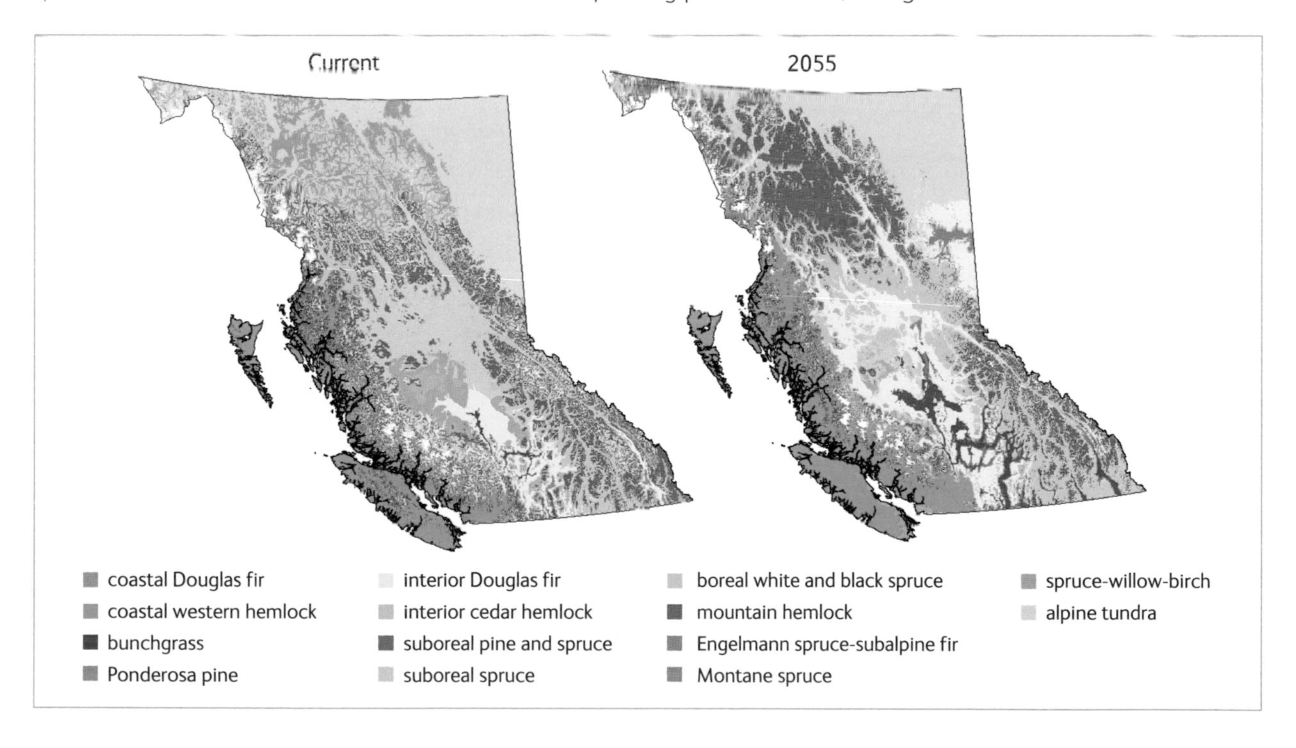

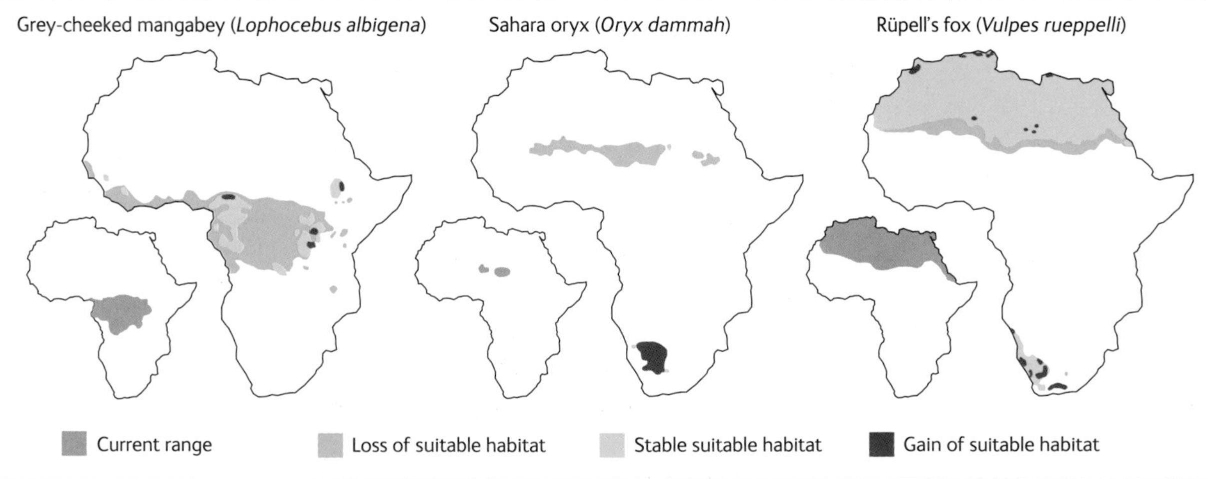
Grey-cheeked mangabey (Lophocebus albigena)
Sahara oryx (Oryx dammah)
Rüpell's fox (Vulpes rueppelli)
Current range
Loss of suitable habitat
Stable suitable habitat
Gain of suitable habitat

these trees grow better, they tend to be more susceptible than local trees to the very occasional severe winter. A warming climate could therefore soon tip the balance in favour of assisted migrants if the worst winters become less severe.

In sub-Saharan Africa, birds may be able to cope with the slower rates of climate change that are expected there, so long as they are given room to move within a network of reserves designated Important Bird Areas. The bird species present within individual reserves are expected to change considerably, but if the reserves are not eroded by development, the network as a whole could protect suitable habitat for all but a handful of the most vulnerable bird species. The predictions are similar for the larger mammals in Africa's national parks. Communities in individual parks are likely to change radically, but how many extinctions take place will depend upon the scope for migration. If there is no dispersal between areas, 10–15% of species could become critically endangered or extinct by 2050, but with good dispersal, the threat is very much reduced.

OPPOSITE FAR LEFT, LEFT AND ABOVE Grey-cheeked mangabey, Saharan oryx and Rüpell's fox.

FAR LEFT Effect of climate change on the distribution of suitable habitat for three species of African mammals (large maps) compared with their current ranges (small maps).

A study of the effects of projected climate change on nearly 3,000 species of birds, mammals and amphibians in the western hemisphere predicted that the largest changes caused by migration and local extinction will be in the faunas of Central America, the Andes mountains and the tundra of the far north. In the future, the animal species present in the most affected regions will be 90% different from those found there today.

These predictions of how climate change will alter the distribution of species and the composition of communities are based on the simplistic idea that the climatic limits that define the boundaries of species' current distributions will move and that species will expand or contract to those new limits. However, climate is not the only factor that determines where species can live. Every species is embedded in a web of interactions with other organisms. When non-native species arrive in a new area without their natural enemies, their populations can explode (see chapter 14). The simplistic climate models tell us that climate change will shake up the kaleidoscope of nature and that new patterns will result, but they do not tell us what will happen when new combinations of species are thrown together in the process and begin to interact with one another, nor what the consequences will be when existing bonds in the web of nature are tugged apart.

A dislocated web

The leafing of oak trees in spring was faithfully recorded by the Marsham family on their Norfolk estate, but they were not the only creatures with an interest in this spring event. The whole food web in an oak wood attends the bursting of leaf buds like a hoard of hungry schoolboys waiting for the lunchtime bell. Fresh, unfolding leaves, still young and tender and unprotected by poisons or a tough waxy cuticle are choice caterpillar food. Eggs of the winter moth, the oak leaf roller and other moths anticipate the emergence of the spring food supply, timing their moment of hatching by the weather cues that trigger budburst in the oaks on whose leafless twigs they have passed the winter. Timing is everything and when the caterpillars get it just right, their voracious appetite for oak leaves can render a whole oak wood leafless. If the caterpillars hatch a little too late, the leaves have already become poor quality food and their consumers grow more slowly.

The caterpillars that devour the spring harvest of oak leaves are in their turn eaten by songbirds such as blue tits, great tits and the pied flycatcher that feed caterpillars to their young. At the top of the food web hovers the sparrowhawk that preys upon small songbirds to feed to its chicks. A study in the Netherlands has found that each of the species in this food web responds to spring temperature in an individual way. Over the last 20 years or so caterpillars' ability to track budburst has not altered, but songbirds are now between 5 and 15 days out of synchrony with the peak availability of caterpillars. The pied flycatcher is the songbird that has become most out of synch with its food because it migrates each spring from Africa, where the climatic cues that tell it when to begin the migration have not kept pace with the advance of spring at its breeding sites in the north of Europe. In places where the mismatch in timing between the arrival of the pied flycatcher and the availability of its caterpillar food is greatest, numbers of this

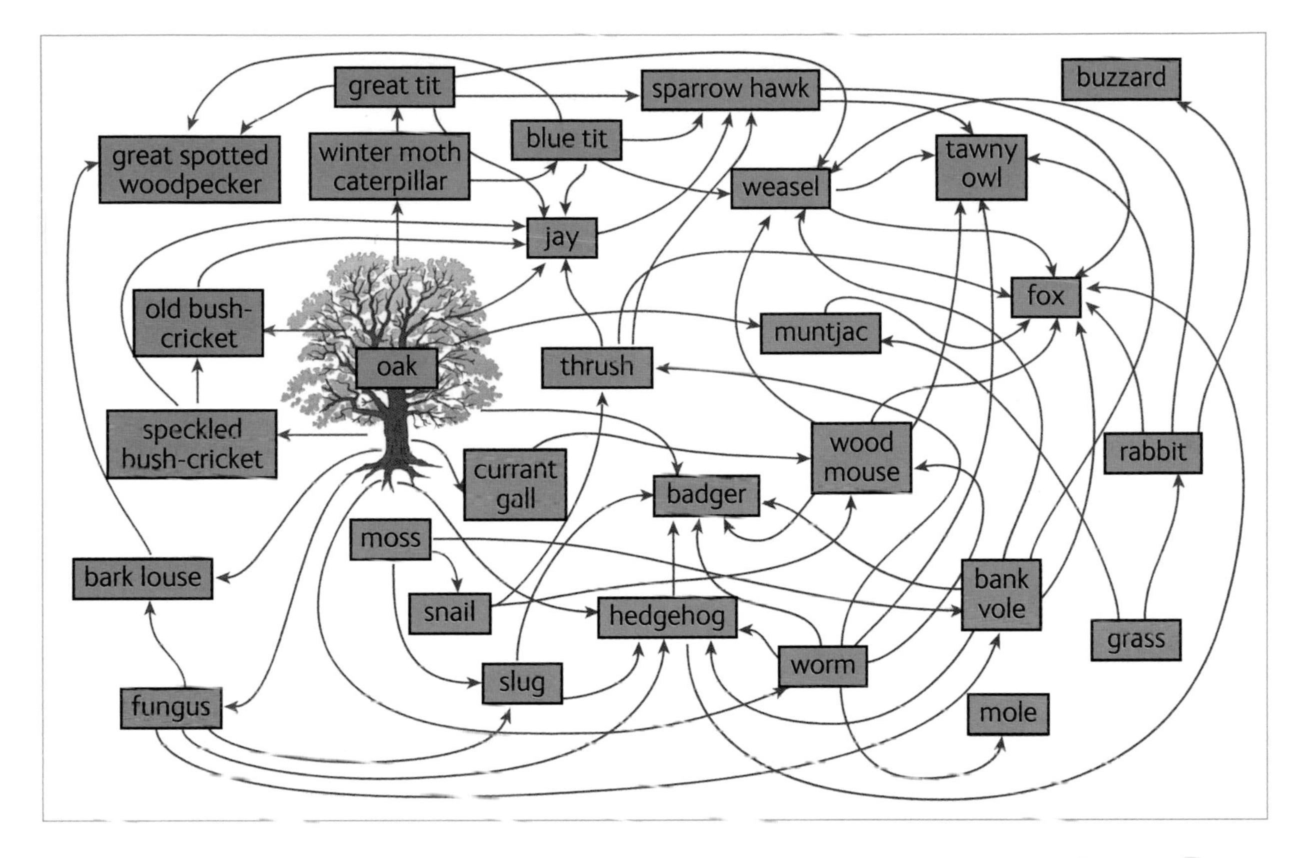

Above The food web of an English oakwood. Arrows connect prey with the species that eat them.

songbird have declined by 90% compared with places where birds and caterpillars are still in synch. At the top of the food web, sparrowhawks are 20 days out of synch with the breeding of their songbird prey, but because sparrowhawks feed upon many species and take older birds as well as chicks, they are probably cushioned from the effects of this mismatch and at the time of writing, no population decline has been detected.

While some of the consequences of climate change, such as the miss-timing of bird reproduction are predictable, and were indeed predicted, how other interactions in the web of nature will change is harder to foretell. Perhaps the ones we should pay closest attention to are those that involve diseases that are transmitted from sufferer to sufferer by another species (called a vector) such as a mosquito or a tick. In 2006, blue tongue virus, which causes a devastating disease of cattle, sheep, goats and deer and was formerly restricted to warmer climates, began to appear in northern Europe. The virus is normally transmitted by a species of mosquito that is found in warmer climates, but the spread of blue tongue virus to northern Europe was not simply a case of this vector moving farther north as the climate warmed, although its range did change. Climatic warming brought the virus into contact with mosquitoes that have a more northern distribution than its normal vector and the virus then adapted to proliferating in the cells of two species that had not previously been hosts. As a consequence, the virus is now present in mosquitoes that are quite at home in the cooler environment of northern Europe. Climate change provided the springboard for this change, but the subsequent northwards spread of blue tongue virus has been the result of the evolution of the virus and its adaptation to two new vectors.

CORAL GRIEF

ABOVE Changes in water temperature cause coral to expel or suppress their algae, which gives the coral a bleached look.

Coral reefs are like the rainforests of the sea, they support so many species. A quarter of known marine animals spend part or all of their lives in or near coral reefs, including 4,000 species of fish. Their value to tourism and fishing just in Hawaii and Florida is $2 billion a year. The Millennium Ecosystem Assessment valued all the ecosystem services provided by coral reefs at $30 billion a year. A worldwide evaluation of reef building corals published in 2008 found that in the previous decade there had been a huge deterioration, with hundreds of species becoming threatened or near-threatened. Fully one-third of coral species are now at an elevated risk of extinction from a variety of causes including coastal pollution, over-fishing, coral mining and rising atmospheric CO_2.

The rise in atmospheric CO_2 presents a dual threat to corals. The first is through rising sea temperatures caused by global warming. Most reef-building corals in warm surface waters are a symbiosis between colonial animals that belong to the same group as jellyfish and photosynthetic single-celled algae called zooxanthellae. Corals are very sensitive to water temperature and respond to sudden changes by expelling or suppressing their zooxanthellae, which gives the coral a bleached appearance and eliminates much of their food supply. If normal conditions are restored again, zooxanthellae often re-colonize. If they do not, most corals eventually die. How resilient a coral reef is to disturbance depends, among other things, upon how intact its fish fauna is. Herbivorous fish graze encrusting algae, allowing light to get through to the living coral whose zooxanthellae require light for photosynthesis. If fish are removed, the algal growth is unchecked and the whole reef changes irreversibly from supporting a rich variety of species to a much impoverished mass of dead coral. Sediment and pollution washed off adjacent land can kill coastal reefs in much the same way.

The second threat that the rising level of CO_2 in the atmosphere presents to corals is so serious that many coral experts forecast that reefs, such as the Great Barrier Reef off the coast of Australia, will be completely dead by 2050. This threat is ocean acidification, caused by CO_2 dissolving in sea water. As the concentration of atmospheric CO_2 rises, the ocean becomes more acid and this inhibits the ability of corals to grow their calcareous skeleton, essential to keeping near the warm, well-aerated, well lit surface waters.

ABOVE The Great Barrier Reef in Australia is one of those under threat from global change in temperature and atmospheric CO_2.

Climate change is occurring at such a rapid rate that most animals and plants are unable to evolve fast enough to keep up but, as the example of blue tongue virus shows, microbes, which can multiply very rapidly, can and will evolve in response. Such fast evolution can be to the advantage of some of the plants and animals like corals that depend upon them. A study of the genetics of zooxanthellae, the single-celled algae that live in corals (*see* opposite), before and after a bleaching event in the Great Barrier Reef in Australia found that the original zooxanthellae belonged predominantly to a temperature-sensitive form, but 6 months after the event the commonest form was tolerant of high temperature. Models suggest that genetic variation among zooxanthellae could help corals to adapt to climate warming, so long as the change was not too great. Whether this will make any difference to the fate of corals in the long run depends upon the other threats they face.

Another obligate symbiosis occurs between aphids (e.g. greenfly and blackfly) and a bacterium called *Buchnera*, which the insects harbour within the body. Aphids live on plant sap that is very high in sugars but practically devoid of the amino acids that are needed to make proteins. The *Buchnera* bacteria provide aphids with essential amino acids and without their symbionts aphids are unable to reproduce. Experiments with pea aphids have found that tolerance of these insects to a heat shock depends upon a gene carried by their bacterial partner. With bacteria carrying the right allele, aphids are not affected by a heat shock, but if the bacteria have the cool-adapted allele, the bacteria are knocked out and aphids do not reproduce. The existence of the alternative alleles therefore permits both *Buchnera* and its aphid host to adapt to climate change. Climate change is a major factor in the future of nature, but so too is globalization and the movement of species around the planet.

14 CHAPTER *Invasion planet Earth*

IN WILLIAM SHAKESPEARE'S PLAY, *HENRY THE FOURTH (PART 1)* Hotspur, a nobleman at odds with King Henry boasts that 'I'll have a starling shall be taught to speak,' in order that the bird should incessantly taunt the King with the name of a rival to the throne of England. From this casual line, written for the dramatic effect it would have on a London stage at the end of the sixteenth century, arose a plague of starlings 300 years later and 1864 km (3,000 miles) away. The culprit for the introduction of the European starling into North America was a wealthy New Yorker named Eugene Schieffelin who resolved to combine his love of birds and the Bard by introducing into the USA every species mentioned in Shakespeare's works. After failing with introductions of several species, Mr Schieffelin imported 80 starlings from Europe that he released into Central Park, New York City. For good measure he released 40 more a year later and by 1904 starlings had multiplied and spread to Connecticut, New Jersey and upstate New York. Now, the millions of descendants of these birds occur from coast-to coast in the USA and also in Mexico, Bermuda and Canada. The European starling has been introduced into other continents and now has an almost global distribution.

The European starling is an aggressive bird that usurps the nesting holes used by other species. Surprisingly, despite this behaviour, a study in 2003 that compared bird numbers before and after starlings arrived at various locations in the USA could not find much evidence that starlings had actually caused declines in native species of hole-nesting birds. Consequently, starlings may not be guilty of displacing native birds in areas that they have recently invaded, a rare mercy as non-native species often cause havoc for indigenous biodiversity. In the UK, where the starling is a native and where it was until recently very abundant, there has been a dramatic decline in the species. The decline in starlings in the UK has coincided with a rise in the abundance of another hole-nesting bird, the great spotted woodpecker, which seems to have benefited from the reduced competition for nest holes.

Many of the European species that are troublesome in North America were deliberately introduced or spread by 'acclimatization societies' formed by people like Mr Schieffelin who believed that they were improving the biological riches of the New World. Partly for this reason, the traffic in invasive species across the Atlantic has not been symmetrical, with many more European species following the path of human migration from Europe to North America, than have travelled in the reverse direction. Charles Darwin even ribbed his American friend Asa Gray about it in a letter in which he asked 'Does it not hurt your Yankee pride, that we thrash you so confoundedly? I am sure Mrs Gray will stick up for your own weeds. Ask her whether they are not more honest, downright sort of weeds.' To which Mrs Gray replied that American weeds were

OPPOSITE Starlings are native to Europe, but are now present globally thanks to introductions and their huge capacity to increase.

OPPOSITE A stand of paperbark being sprayed with herbicide to control this invasive species in the Florida Everglades.

'modest, woodland, retiring things; and no match for the intrusive, pretentious, self-asserting foreigners.'

If the European weeds have an excuse, it would be that many of them are merely doing in North America what comes naturally to them in Europe, colonizing the facsimiles of European agricultural habitats that have been re-created in North America. For example, a whole émigré community of annual plant species has gathered in the Mediterranean climate of California to form a new type of grassland vegetation that could not have existed there before European settlement. The native vegetation in these areas that was probably composed of perennial grasses disappeared before any botanist was able to record it. The introduction of European grazing animals to California may have helped the annuals replace the native plant species.

ABOVE Purple loosestrife is a highly invasive plant in North America, but rarely a problem where it is native in Europe.

One plant invader of North America that cannot be said to be behaving as it does at home is purple loosestrife. In Europe, this plant can become locally or temporarily dominant in the wet places where it grows, but it does not get out of hand. In North America purple loosestrife has conquered millions of hectares of wetlands, excluding native vegetation and pushing some rarer native plants to the brink of extinction. Purple loosestrife has a profusion of attractive purple flowers which undoubtedly helped it obtain a ticket to travel from Europe. The very invasive mile-a-minute-vine, or kudzu, was also introduced from Asia because of its attractive purple flowers. Then it was deliberately spread and planted to control soil erosion in the exhausted cotton fields of the US South where it was recently estimated to cover 809,371 ha (2 million acres) of forest land alone.

There are dozens of other seriously pestilential plants in North America that have been deliberately introduced for various misguided ends. Florida, with its semi-tropical climate, is especially benighted by invasive plants, ceding the trophy only to Hawaii as the most invaded state in the Union. One of the major problem plants in the Florida Everglades is an Australian tree called paperbark that was deliberately sown by scattering seeds from aircraft during the 1930s. The aim was to use paperbark, which is much thirstier than Florida's native trees, to drain areas of the Everglades for use in agriculture. The result was millions of acres of dense, wall-to-wall paperbark forest in which nothing except non-native plants would grow. Paperbark seeds are adapted to survive fire and when stands of the tree are burned, they simply regenerate from seed in even higher densities than before.

My enemy's enemy

For anyone interested in keeping score in the international contest for most biologically invaded nation, Australia plays in a league of its own. Ironically enough, one of its problem plants in wetland areas is a tree called pond apple that is a native of the Everglades. Back home in Florida, pond apple is threatened by Australian paperbark invading its habitat. This bizarre situation, in which each species does better in away territory than on its home turf, is in fact only a reciprocal example of a pattern that is quite common among non-native species. Invading species often do better in territories where they are not native. The recent decline of the starling in the UK makes this species one of the latest additions to this list. The likely reason for this counter-intuitive phenomenon is instructive because it suggests how non-native species can be controlled.

Plants often travel abroad as seeds or small specimens that do not harbour the many insects, fungi and viruses that afflict them in their native range. Most natural enemies, including fungi and many herbivorous insects, tend to be specialized and attack only a narrow range of species. As a consequence, plants from elsewhere that have become naturalized in the USA are attacked by only a small fraction of the species of rust, smut and powdery mildew that infect them in their native ranges. This emancipation means that when native and non-native plants come into competition with one another, the natives are handicapped by their natural enemies while the non-natives are largely unmolested. This mechanism explains why non-natives often do better in new lands than they do back home, but it also points to an obvious remedy for invasions – import the natural enemies of invasive species. This strategy of biological control can be spectacularly successful if it is done with proper care.

The natural enemies used in biological control have to be carefully screened before introduction to ensure that they will only attack the desired target and not become pests themselves. Introducing general predators can lead to an escalation of problems, as the *Old Woman Who Swallowed a Fly* discovered in the last verse of the children's song:

There was an old woman who swallowed a cow,
I don't know how she swallowed a cow!
She swallowed the cow to catch the goat,
She swallowed the goat to catch the dog,
She swallowed the dog to catch the cat,
She swallowed the cat to catch the bird,
She swallowed the bird to catch the spider,
That wriggled and jiggled and tickled inside her,
She swallowed the spider to catch the fly,
I don't know why she swallowed the fly,
Perhaps she'll die.
There was an old woman who swallowed a horse,
She's dead – of course!

OPPOSITE A lake in Kenya infested with the non-native water hyacinth before and six months after the introduction of a weevil that lives on the plant in its native South American region, and which successfully cleared the lake.

Several situations reminiscent of *The Old Woman Who Swallowed a Fly* have occurred in Australia. Indian mynah birds were imported to control locusts and became a pest,

followed by Indian mongooses to control rats. Then Australia swallowed the giant cane toad to catch a beetle that attacks sugar cane. Within 6 days of the cane toads reaching Australia they had laid their first eggs in the pond that had been constructed to welcome them. The following year over 41,000 baby toads were distributed to eager cane farmers and the breeding pond was continually re-filled with hundreds of thousands of eggs by its inhabitants. In the event, the giant toads did not eat the cane beetle because the beetle larvae were hidden inside cane stems, but they ate other insects including large numbers of the dung beetles that had been introduced to deal with Australia's cow dung problem (*see* chapter 6). Cane toads are highly poisonous. Native predators poisoned by them have been pushed onto the list of endangered species as they encountered the advancing front of the cane toad invasion. Populations of freshwater crocodiles have crashed by 77% following exposure to cane toads. Research is currently being conducted on biological control of the cane toad using some of the natural enemies it left behind in its native South America.

Though invading plants and animals benefit from leaving their natural enemies back home, they are not always disease-free. Some invading animals carry diseases to which they are immune but to which indigenous species are susceptible. Over many decades, animal-lovers in the UK have watched their beloved native red squirrel retreat before the remorseless advance of the American grey squirrel, which has replaced it in many areas. The two species do not interact in the way that aggressive starlings do with other hole-nesting birds and so it was not immediately clear why grey squirrels were replacing

BELOW Cane toads were introduced to Australia to catch a beetle that attacks sugar cane, but were unable to reach the larvae hidden inside the plants stems so ate other insects instead. Some native predators, including freshwater crocodiles, have been poisoned by the toad and as a result have become an endangered species.

red. The answer turned out to be that many grey squirrels carry a pox virus that is fatal to the reds. An even more devastating non-native in Europe is the American signal crayfish. This species carries a fungal infection called crayfish plague which, along with its aggressive and predatory behaviour, has caused it to replace native white-clawed crayfish in many rivers throughout Europe. Exotic bird species brought avian malaria to Hawaii and the disease has exterminated almost all native birds below an altitude of 1,700 m (5,600 ft) where the mosquito that transmits malaria lives. Non-native plants carry disease too, chestnut blight being a devastating example that arrived in North America with introduced trees (*see* chapter 7).

Extinction by association

Invasion by non-natives is just one of many environmental challenges to the survival of indigenous species. Non-natives often invade environments that have already been altered or created by humans, leading some people to challenge whether invasions are really the cause or just the symptom of a problem. Each case needs to be examined on its merits but there are many examples where non-native species have been clearly established to be the cause of extinction and a very much bigger number where natives are demonstrably threatened by competition, predation or disease associated with non-natives. The American signal crayfish presents all three of these types of threat to native European crayfish.

The majority of global extinctions so far recorded have been on islands (*see* chapter 11) where introduced rats have driven many birds to extinction. Other mammalian predators can also be a threat to ground-nesting birds. Hedgehogs were introduced into the Scottish Hebrides in the 1970s where they multiplied enormously, living on the eggs of dunlin, corncrake, lapwing, skylark, redshank and snipe. Hedgehogs are now being trapped in the Hebrides and released on the mainland where they are native and their numbers are naturally controlled. Control of non-native animals is possible even over large areas if it is done systematically. Rats have been successfully cleared from even quite large islands. Coypu, a large South American rodent that colonized parts of East Anglia and damaged the native vegetation, was successfully eradicated in less than 10 years once a plan based upon sound scientific research was implemented. Foxes and feral cats have been successfully reduced over an area of 3.5 million ha (8.6 million acres) in Western Australia, allowing indigenous species that were formerly threatened to recover. This control programme ingeniously uses bait laced with a poison produced by indigenous plants. Indigenous animals are resistant to the poison, but non-native predators are not.

Freshwater habitats suffer many impacts that threaten their fauna, but of 40 recorded extinctions of North American fish species, 27 are attributable to the impact of non-native species. The most threatened of all freshwater animals in North America are the endemic freshwater pearly mussels. Of the 300 or so species that have been described, 12% are presumed extinct and another 60% are endangered or threatened. Their river habitats have long been damaged by pollution and populations have been over-harvested but a new threat appeared in the 1980s when the zebra mussel appeared in North America (*see* p.162).

THE ZEBRA MUSSEL INVASION OF NORTH AMERICAN LAKES AND RIVERS

This small mussel, no bigger than the nail on your little finger, multiplies to enormous numbers and filters phytoplankton from the water so effectively that these huge populations can starve native mussels and zooplankton. The zebra mussels settle on all available hard surfaces, including the shells of other mussels, and can choke even large water pipes with their numbers. The cost to North American power-generating plants and water-treatment facilities between 1989 and 2004 alone was $267 million. Zebra mussels are also spreading in Europe. A study of the impact of the zebra mussel on the ecology of the River Hudson in New York found that by filtering out phytoplankton, the mussels changed the food web from one containing sport fish that live in the water column (the pelagic zone) to a food web based on submersed plants and fish that feed along the shoreline (the littoral zone). Native, pearly mussels were reduced to a tenth of their densities before the invasion. Whether the indigenous species in the Hudson will be able to survive or recover from these lower densities remains to be seen, but almost all 24 species of pearly mussel that were present in the first North American river to be invaded by zebra mussels are now extinct there.

BELOW Invasive zebra mussels profoundly alter aquatic ecosystems.

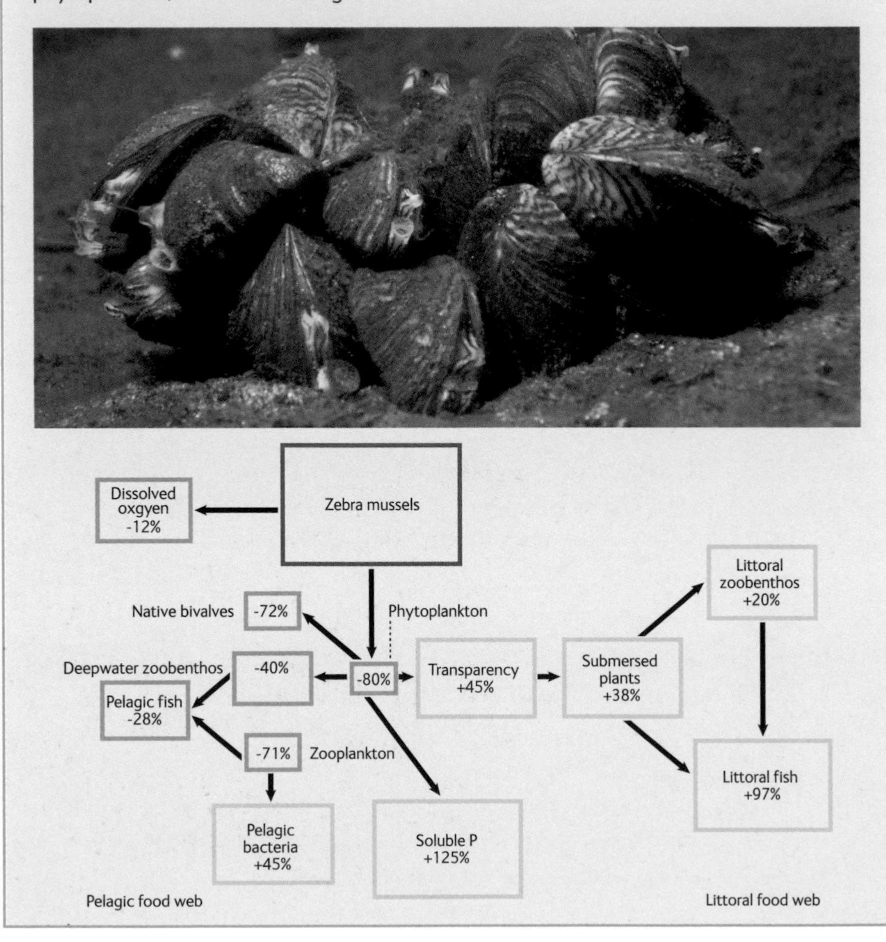

Ecosystem engineers from hell

Invasive species can re-engineer ecosystems, alter food webs and thereby threaten indigenous species. Zebra mussels are not the only ecosystem engineers from hell. The Argentine ant is a terrestrial invader that is now nearly global in its distribution. The species forms extraordinary super-colonies of thousands of genetically identical ants in nests that spread over kilometres. Within these super-colonies, the ants waste no time warring with their kin and instead devote all their energies to attacking indigenous insects including other ant species. The toll of the Argentine ant's effects upon food webs includes driving away pollinators from flowers in the Cape Floristic Region; excluding all native ants including those that disperse seeds in the south of Spain; and changing the indigenous community of ants and other ground-dwelling insects in south Australia and many other places that are warm enough for it to invade.

Plants are well represented among the diabolical ecosystem engineers. In Tahiti, invading miconia trees form solid stands that cover two-thirds of the island, threatening 40–50 species of indigenous forest-floor plants with their impenetrable shade and causing soil erosion as a result. In the Galapagos archipelago, guava and quinine trees have invaded more than 4,450 ha (11,000 acres) of formerly treeless habitat on Santa Cruz Island, threatening indigenous plants that are not tolerant of shade.

LEFT Invasive miconia trees create impenetrable shade and cause soil erosion and landslides in Tahiti.

ABOVE *Caulerpa* is a very damaging invasive seaweed in the Mediterranean.

Such ecosystem-altering effects of invasive species are not confined to the land. One of the worst marine examples is the tropical alga *Caulerpa* that was dumped into the Mediterranean when the public aquarium in Monaco cleared out its displays. It took only 15 years for this alga to carpet 4,000 ha (10,000 acres) of sea floor around the coasts of France, Spain, Italy and Croatia. Fed by nutrient pollution from the land, *Caulerpa* forms densely packed underwater swards with as many as 14,000 fronds per square metre (12,500 per square yard). Fish are unable to penetrate such a closely woven carpet to feed upon animals living in the mud. *Caulerpa* has replaced indigenous sea grasses on which molluscs, sea urchins and herbivorous fish used to feed, but the invading alga is toxic and so the whole food web that depended upon sea grasses has been undermined.

Wildfire is a natural event in many ecosystems, particularly grasslands, shrublands and certain types of forest. The animals and plants that live in such environments are adapted in various ways to survive or recover after fire, but these strategies only work if fire is not too frequent. Pines, for example, have to be able to grow large enough to accumulate an aerial bank of seeds in their canopy. Seeds are sealed inside resinous cones until the heat of a fire releases them to fall into the newly cleared ground. If fires are too frequent, trees do not have time to build up their seed bank and there may be insufficient seeds to replace the pines that die in the fire. Invasive grasses often alter fire risk to the detriment of indigenous species. In the Sonoran Desert in the American Southwest, the frequency and extent of summer fires have been increased by the invasion of African and Mediterranean grasses. Desert tortoises and saguaro cacti are endangered as a result.

There is a long list of 'usual suspects' among invasive species, like the starling, the zebra mussel, the Argentine ant, the black rat, *Caulerpa* seaweed and water hyacinth, that seem to spell ecological trouble wherever they go, except of course in their own native environments where their dominating tendencies are controlled or accommodated. These global species are inexorably squeezing out the indigenous species that make one place different from another. Managing invasive species is now an essential part of the attempt to conserve nature everywhere. The onward march of invasive species, climate change and the increasingly weighty impact of our own footprint upon nature all prompt the urgent question: what next for nature?

15 *What next for nature?*

CHAPTER

FOR GOOD OR ILL, THE FUTURE OF NATURE IS NOW LARGELY IN OUR HANDS. Our activities are now so all-pervasive that we live in a new geological epoch where global processes are shaped by humans, an epoch that the Nobel Prize winner Paul Crutzen dubbed the *Anthropocene*. Whether we directly hunt species to extinction, inadvertently drive them in that direction by altering the climate or, conversely, spread our favourites to every corner of the globe, it is difficult to avoid the conclusion that, like it or not, we are responsible for the fate of nature. This situation has been a long time coming, but the signs have been present for a least a millennium and probably forty, ever since our species reached Australia and many of the largest animals there went extinct (*see* chapter 11). Big animals have low birth rates that make them particularly susceptible to extinction when hunted. Direct human responsibility for these ancient extinctions of megafauna has not been conclusively proved, but we have been repeatedly found at the scene of the crime holstering a smoking gun.

The whodunnit of megafaunal extinction has a curious twist and a fateful sequel. The twist is that while human migration out of Africa left a wave of extinction in its wake, many giant mammals and a flightless bird (the ostrich) survived in the ancestral home of the human species. Why the megafauna survived better in Africa of all places

OPPOSITE The ostrich is the last surviving species of giant flightless bird.

BELOW Rhino and lion are members of a megafauna that survived in Africa, while large mammals and birds went extinct elsewhere.

is a conundrum. It is tempting to draw a parallel between the relative harmlessness to wildlife of humans in our place of origin with the stable ecology of many invasive species within their native ranges (*see* chapter 14). The fateful sequel to this story is of course that the African megafauna, as well as many of the larger animals that remain on other continents, such as the tiger, the giant panda and the brown bear, are now threatened. Although this sequel seems ominously familiar, its outcome can and should be different. This time we are the knowing authors of nature's fate. So, what should we do?

Use it or lose it?

The key to protecting biodiversity is the buzzword of environmentalism – *sustainability*. What it means in the context of biodiversity is that exploitation of a species should be limited to a level that does not endanger the long-term viability of its population (*see* opposite). How can the exploitation of a species be limited to sustainable levels? Biodiversity has value and one approach to conserving species is to calculate their explicit monetary value and then to rely upon the market to ensure that a sustainable supply of the valuable commodity is preserved. Placing an actual value on biodiversity and ecosystem services is not straightforward. An attempt to quantify the value of ecosystem services, including the harvests of food and other products that we take from biodiversity on land and in the sea, came up with the figure of $33 trillion (thousand billion) per year, or nearly twice global gross national product ($18 trillion) for the late 1990s when the estimate was made. The difference between the two figures (i.e. $15 trillion) could be interpreted as a very approximate measure of how far the value of ecosystem services and biodiversity is taken for granted – a lot! However, since human existence would be impossible without ecosystems, their true value is literally incalculable, in which case $33 trillion could be regarded as a very low estimate of infinity.

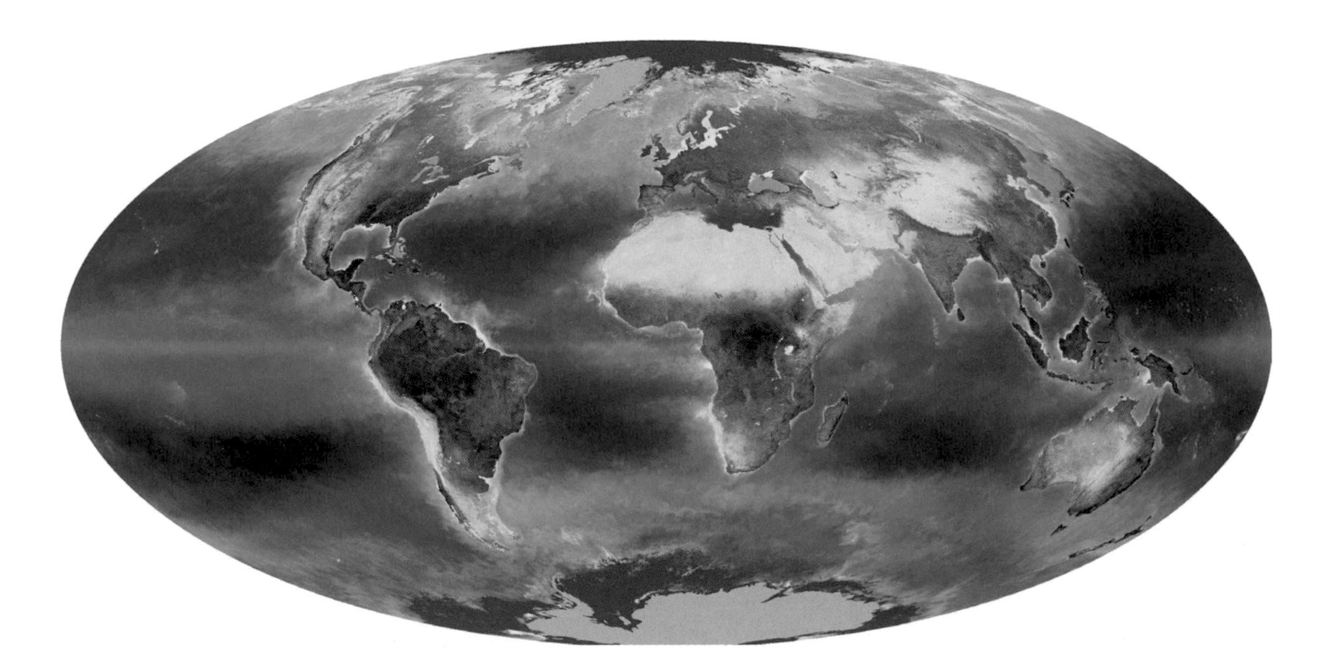

BELOW Global distribution of plants on land (darker green indicates more plant mass) and phytoplankton in the oceans (lighter blue and green indicate areas of greatest concentration).

SUSTAINABLE EXPLOITATION OF BIODIVERSITY

Crudely, if its death rate exceeds its birth rate for a substantial period, a population is in trouble. If the birth rate exceeds the death rate, the population may be able to withstand some harvesting – effectively, an increase in the death rate. Of course nothing is quite so simple, chiefly because there is always uncertainty about what death rates and birth rates actually are and about how they may change in the future. A safety margin should be maintained, but how big is necessary? There should also be a floor of minimum population size below which no population should be allowed to go, even if its birth rate seems healthy. Small populations are inherently vulnerable to extinction, at least in part because they are often genetically depauperate, which can be a hazard for long-term survival and adaptability (see chapter 5). Population size is one of the criteria used by IUCN to decide whether a species should be listed as threatened (see chapter 11).

Once the economic value of biodiversity is properly appreciated, why, one might naively ask, would anyone kill a goose that lays such an abundance of golden eggs? Using an explicit value for ecosystem services in the hope that the hidden hand of market forces will protect the biodiversity upon which ecosystem services depend might be called the *use-it-or-lose-it* conservation strategy. In an ideal world, and with a perfect market for gold eggs, their value alone might be sufficient to protect the goose that lays them, but several real-world problems arise. The first is related to ownership and inequality. If you are rich and own a flock of geese and I am poor and my children are starving, I might well kill your goose if that is the only way I can acquire some golden eggs, or eat. Poaching of wildlife is a major problem in many poor countries for this reason.

Even if the goose is common property and belongs to everyone (or to no one), it is in my selfish interest to take what I can get rather than to share. This so-called tragedy of the commons can lead to environmental degradation through practices such as over-grazing of livestock on common land and over-fishing of the seas, the problems of an unregulated (or poorly policed) system of harvesting wild populations. Giving users a long-term stake in the exploitation of a resource and its management can reduce the incentive to over-exploit it. Such a strategy has helped to restore inland fisheries in Bangladesh (*see* chapter 11).

Regulated harvesting can be problematic too. You might be persuaded that you can afford to kill a few of your own geese because you think that you have quite enough to sustain a population into the foreseeable future and anyway, you need that goose now and the future can take care of itself. Fisheries have followed this pattern into one collapse after another. For political reasons, current fishing quotas are often set at high levels that fishery scientists say are not sustainable (*see* chapter 12).

Finally, there is the problem that if you put your faith in markets to protect species, then biodiversity becomes a commodity just like any other and investors could quite rationally decide that the best financial strategy is to hunt geese to extinction and then invest the profits in some other industry. If the bank pays you 5% on your money and goose hunting pays only 4.5%, then the logic of the market says you should cash in all the geese you can get straight away and put the money in the bank. For populations of large slow-growing trees or large animals that recover only slowly from harvesting, like whales for example, the financial rate of return on harvesting them sustainably is always likely to be kept low by the biological fact that big organisms have low birth rates. These most vulnerable and charismatic of species could be doomed by exposure to an unfettered market in their products. Many of these issues are illustrated by the successes and failures of conservation strategies for the most mega of the African megafauna, the elephant.

AFRICAN ELEPHANTS

In 2001, it was discovered that forest elephants and savannah elephants in Africa were so genetically distinct from each other that they were two distinct species. The African forest elephant, *Loxodonta cyclotis* (below), occurs in Central and West Africa. Forest animals are difficult to census and war conditions in parts of the region make population estimates harder to obtain, but the evidence suggests that this species is particularly threatened. The African savannah elephant, *Loxodonta africana* (right), occurs in forest and grassland environments in east and southern parts of the continent. As you would expect of six-ton herbivores that live in herds and can eat woody vegetation too tough for other animals, elephants have a dominating effect upon ecosystems. In lower rainfall areas, elephants can convert savannah vegetation with scattered trees to grassland. But elephants have a variety of beneficial effects; they disperse seeds in their dung, which is important in recycling nutrients, they dig water holes used by other species, and movement through vegetation disturbs insects and small animals that birds feed upon. From a conservation point of view, a nature reserve big enough to sustain elephants is also large enough to protect a

Tackling the ivories

Elephants are an excellent test case for the use-it-or-lose-it strategy because elephant ivory is a luxury product with a high price. Rising affluence in Japan in the early 1970s triggered an increase in demand for ivory that was met by a surge in elephant killing. A convention regulating international trade in endangered species (CITES, *see* p.173) was introduced in 1975 and this attempted to regulate trade in ivory by requiring a permit. However, mainly as a result of poaching, the total population of African elephants fell by more than a half in the ensuing decade. What ought to have been a case of use it

huge range of other species. Elephant impact extends to humans too, of course, and problems tend to arise at the boundaries of reserves where elephants raid farmers' crops and threaten their livelihoods. But, the ecological web provides an imaginative answer to this problem. Trials have found that African elephants will stay away from a fence line that is guarded by African honeybees, which are more aggressive than European bees, with a sting powerful enough to deter pachyderms. The initial cost to the farmer of providing hives can be defrayed by the value of the honey that the bees produce.

LEFT AND BELOW African forest, *Loxodonta cyclotis*, and savannah elephants, *Loxodonta africana*, are distinct species, although they do commonly hybridize.

ABOVE Burning illegal ivory in Kenya after elephants were moved to Appendix I of CITES.

or lose it proved in actuality to be a case of use it *and* lose it. Having failed to regulate the market successfully, the next idea was to abolish the international market entirely and so remove the incentive for poachers to kill elephants by lowering the international price of ivory to zero. A complete ban on international trade in ivory was introduced in 1989. The government of Kenya advertised the ban by publicly burning 2,000 elephant tusks that they had confiscated from poachers. Purchasing ivory products became unfashionable and the price of ivory fell drastically.

The listing of African elephants on Appendix I of CITES, which bans all international trade, was opposed by countries in southern Africa that argued that poaching was not a serious threat to elephants in their well-managed game parks. They argued that the use-it-or-lose-it strategy was working for them. In retaliation for what was perceived as interference from northern hemisphere countries, there was even talk of listing the herring, at the time a fish in drastic decline in the North Sea, on CITES Appendix I. A compromise that allowed trade in ivory from some countries but not others was at first resisted because of the fear that the legal trade would be used as a cover for an illegal market in poached ivory. However, in 1997 African elephants from populations in Botswana, Namibia and Zimbabwe were transferred to Appendix II and the one-off sale of 50 tons of raw ivory to Japanese traders was permitted as an experiment in 1999. The management of elephants is still a highly contentious issue. The ivory ban

CITES – A CONVENTIONAL APPROACH TO SAVING WILDLIFE

Some endangered species have huge commercial value, for example in the market for Chinese medicine, which uses many animal parts such as tiger bones, in the pet trade and for luxury goods such as crocodile-skin handbags. The Convention on International Trade in Endangered Species of Wild Flora and Fauna (CITES) came into effect in 1975 in an attempt to protect some 30,000 species of endangered animals and plants that may be threatened by harvesting for international trade. The species affected by CITES are listed in three appendices to the Convention which impose different levels of restriction on trade. The most endangered species are listed in Appendix I; they may not be imported into any country except for limited non-commercial purposes that require a special permit from the country of origin. The tiger was listed in Appendix I from its inception and the African elephant was initially given the lesser protection of listing on Appendix II. Appendix II allows limited trade in less severely threatened species, but only under a permit that can only be granted if trade is not detrimental to the survival of the species in the wild. Listing of species on Appendix III is used by countries to request the help of others in controlling the trade in species such as tropical trees with high-value timber that are being harvested unsustainably or exported illegally.

BELOW A store of confiscated ivory.

did lead to a partial recovery in the global population of African elephants because of its success in southern and eastern Africa, although extensive poaching continued in Central and West Africa where a decline has continued. There is no direct evidence that the licensed sale of limited amounts of ivory, in 1999 and subsequently, has offered cover for sales of poached ivory. Elephant poaching has increased, but the illegal ivory collected is smuggled and not passed off as legal. Twelve seizures of contraband African ivory destined for the Far East made in 2005–2006 contained over 23,000 kg (50,600 lb) of fresh ivory. At least 2,300 elephants must have been killed to obtain this amount. Of course exact figures for ivory smuggling are impossible to obtain, but if the 12 seized shipments represented 10% of the contraband ivory shipped that year (which is the proportion of contraband drugs that customs officers estimate that they intercept), then 23,000 elephants were killed by poachers for the international trade in that year alone.

It appears that killing of African elephants by poachers may be undergoing a sharp increase, driven once again by increasing demand from the Far East, this time by growing affluence in China. In 2002 Zambia requested permission to sell ivory from its stockpile, estimating in support of the application that only 135 elephants had been poached in that country in the previous decade. Just afterwards, an illegal ivory shipment was seized in Singapore and DNA forensic tests (*see* chapter 5) showed that it contained ivory from between 3,000 and 6,500 Zambian elephants. The discovery caused Zambia to replace its director of wildlife and to increase its anti-poaching efforts. The market in ivory continues to grow with increasing demand from the Far East where in 2007 the price was $750 per kilo ($340/lb). Ironically, this huge price may be being driven by speculators buying up ivory in the expectation of a future ban on the trade or, more grimly, of the extinction of elephants in all unprotected populations. This attitude of cash-in-while-you-can could doom the elephant. The good news is that protection can work and there has been a significant increase in elephant numbers within reserves in southern and eastern Africa.

The African elephant was a flagship species for CITES and the idea that controlled trade in species on Appendix II might benefit wildlife. This policy has been a success in parts of Africa where poaching can be controlled, though not elsewhere. Elephants are a major tourist attraction to foreign visitors and the value of tourism to many African governments is clearly an incentive to protect the animals from poachers as well as providing the funds with which to pay for this protection. More generally, a recent study that compared projects employing an ecosystem services approach to biodiversity with a more traditional, purely conservation approach, found that the ecosystem services projects were much better funded and no less likely than the other projects to include or create protected areas for nature. The question remains, however, where should the balance lie between use and protection?

Sparing land for nature

The human population continues to rise, especially (though not only) in countries that are rich in biodiversity, but economically poor. There will be at least another 2 billion mouths

to feed by 2050 (*see* chapter 1). Given the demands that population increase will place upon agriculture and fisheries, how much land or ocean can be spared for nature, or will every hectare have to earn its keep through ecosystem services of one kind or another? As we have seen, protected areas in freshwaters and the ocean can boost production in nearby fisheries (*see* chapters 11 and 12) but this commercial justification for creating protected areas does not apply in the same way on land. On land, most of our food comes from intensive agriculture rather than from harvesting wild populations as it does in the sea. In fact, more and more fish are farmed in coastal waters, so even in the sea, the long-term commercial justification for protected areas could be undermined.

ABOVE Will continuing crop improvement enable us to feed the growing human population through high-intensity agriculture, sparing land for nature?

Crop yield per hectare doubled in the last 40 years of the twentieth century, but this advance has had an environmental cost – it required more water for irrigation, more fertilizers, more pesticides and more fuel. The intensification of agriculture and the conversion of natural habitats to farmland between them constitute the largest threat to biodiversity on land. Among globally threatened bird species, 37% of the threats come from agriculture, but there may be even worse to come because agriculture is responsible for the plight of an even greater fraction of near-threatened bird species.

Despite its directly detrimental effects on wildlife, agricultural intensification has had indirectly beneficial effects too. It is calculated that without the yield increases achieved over the last 40 years by scientific crop-breeding, India would need double the area that it has in agriculture at present and China and the USA would need treble the amount, just to meet current demand. So paradoxically, agricultural intensification could be good for wildlife if it spares land for nature, but are past trends a reliable guide to what may happen in the future? Can yields continue to rise at a sufficient rate to spare land from agricultural development? And even if they can, will further intensification occur without collateral damage to the non-farmed environment? Freshwater resources, for example, are already exploited to their limits in many places (*see* chapter 10).

There is some evidence that recent increases in crop yield in developing countries may have slowed the rate of deforestation but the effects are difficult to detect and may be weak.

Wildlife-friendly farming

An alternative to land sparing through agricultural intensification is wildlife-friendly farming. In the European Union about $2.7 billion is spent annually in subsidies to farmers to encourage farming practices that protect biodiversity. Many European landscapes have been cultivated for millennia (*see* chapter 8) and much of the European flora and fauna lives in semi-natural habitats such as meadows, managed woodlands or heathlands that depend upon low-intensity, traditional agricultural practices for their maintenance. The European Union's agri-environment subsidies aim to indirectly compensate farmers for the lower crop yields from less intensive, more biodiversity-friendly farming. Such subsidies are not usually available in the Developing World, from where the increasing demand for food will mainly come, but even there, farming and biodiversity may be reconciled to some degree. After all, biodiversity is the farmers' friend (*see* chapter 7), especially where expensive inputs such as industrial fertilizers and pesticides cannot be afforded.

The sharp end of the conflict between agricultural development and biodiversity in the developing world is where tropical forest is under threat from clearance for agriculture and timber. Here, subsidies to prevent deforestation might be on offer from developed countries willing to pay for the ecosystem service of carbon sequestration that intact forests represent. Sustainable forestry can also be encouraged by certification timber sold for export to developed countries where there is a demand for sustainably produced wood. One of the biggest timber suppliers that is certified as sustainable by the Forestry Stewardship Council is a collection of community-run enterprises in the Mayan Biosphere Reserve in the north of Guatemala. Here, giving local communities a stake in the sustainable exploitation of the forest cut the rate of deforestation dramatically as soon as the scheme was introduced. Just as in Bangladesh (*see* chapter 11), giving people a stake in their local resources prevented over-exploitation. By contrast, illegal logging continued in the core of the Mayan Biosphere Reserve even though this was zoned exclusively for nature protection.

BELOW Dartmoor, UK, where low-intensity farming maintains a landscape favourable to wildlife.

SHOULD WE EAT LESS MEAT?

Another way to spare land for nature would be to reduce the demand for agricultural produce. Since a large proportion of the grain that is grown is fed to animals and it takes 10 kg (22 lb) of corn to produce 1 kg (2.2 lb) of beef, a diet containing less meat could reduce the human ecological footprint if it was an option chosen by sufficient people in countries where meat consumption is high. Eliminating meat from the diet completely would not help because grazing by animals is required to manage many types of species-rich grasslands. If the market for sheep and cattle reared on such grasslands disappeared, so would their plant diversity. Livestock also graze many areas that are unsuitable for arable farming, but where controlled levels of grazing are compatible with, or even necessary for, the conservation of biodiversity.

BELOW LEFT AND RIGHT Scottish moorland grazed by sheep (left), and bison grazing prairie (right).

Epitaph for the Anthropocene

Viewed with historical detachment, as if by a geologist belonging to some distant and forgetful future, the epitaph for our times could be stark:

> *‘The human epoch that its own creators brazenly dubbed the Anthropocene is marked in the geological record by a mass extinction that overtook the flora and fauna on land and in the oceans too quickly for evolution to respond. Nothing like it had been seen on Earth since the end of the Cretaceous, 65 million years earlier. The burgeoning human population stamped its ecological footprint upon nature with the force and indifference of a colliding asteroid.’*

Whether this description is science fiction or science future is up to us. Having established the facts objectively, we should not view the threatened future with detachment. Ignoring the crisis for biodiversity will be a costly mistake and the price of inaction will place a recurring charge upon the health, wealth and quality of life of future generations. We might well be able to survive with a fraction of the species that currently provide us with ecosystem services of all kinds, but is that a future we would choose for our descendants? If not, we need to check and then reverse the rising carbon emissions that are warming our planet and threaten to dislocate nature, with unpredictable consequences. We must halt deforestation, restore habitats and fisheries, create and protect reserves for nature, find sustainable livelihoods for the poor and feed ourselves without endangering biodiversity. We cannot undo all that we have done, but act we must.

Glossary

Allele a variant of a specific gene. Different alleles belonging to the same gene have slightly different DNA sequences from each other, which may affect how the gene functions but often it does not. The more common types of allele there are in a population, the higher is its genetic diversity.

Anthropocene the geological epoch that began with the Industrial Revolution about 200 years ago, defined by the advent of human domination of Earth processes.

Archaea a group of microbes forming one of the three domains of life. The other two are the bacteria (or eubacteria) and the eukaryotes.

Bacteria/Eubacteria a group of microbes forming one of the three domains of life. The other two are the archaea and the eukaryotes.

Biodiversity life's variety at all levels from genetics through to ecosystems.

Biodiversity hotspot as defined by the organization Conservation International this is an area that contains at least 1,500 endemic plant species and that has lost at least 70% of its original habitat.

Carbon cycle the circular route by which carbon is transferred between the atmosphere, plants, animals, soil and ocean sediments and back again.

Carbon sequestration the locking-away of carbon removed from the atmosphere in photosynthesis by its incorporation into wood or other organic materials that are slow to decompose.

Carrying capacity the maximum size of population that environmental resources can sustain without collapsing.

CITES the Convention on International Trade in Endangered Species of Wild Flora and Fauna (CITES) came into effect in 1975 to protect species of endangered animals and plants that may be threatened by harvesting for international trade.

Collapse when applied to a fishery is defined as a decline to below 10% of the peak catch for a particular species or population of fish.

Critically Endangered a species is deemed Critically Endangered by IUCN when the best available evidence indicates that it is facing an extremely high risk of extinction in the wild.

Ecological web *see* Web of nature.

Ecosystem the organisms in a food web and their physical environment collectively. An ecosystem is more-or-less self-contained but may be of any size, from a garden pot containing a plant and some soil to an entire forest or an ocean.

Ecosystem service Ecosystem services are the benefits that ecosystems, and the biodiversity they contain, contribute to humanity. They may be divided into provisioning services (the production of products such as food), regulating services (such as the role of ecosystems in the carbon cycle), aesthetic and cultural (such as the enjoyment we get from nature) and supporting services that underpin the other three.

Endangered a species is deemed Endangered by IUCN when the best available evidence indicates that it is facing a very high risk of extinction in the wild.

Endemic species species that occur only in the area under consideration

Eukaryotes one of the three domains of life. Members are distinguished by having their genetic material contained within a nucleus in the cell. Many, but not all, eukaryotes are multicellular. The other two domains are the microbial archaea and bacteria.

Extinct a species is deemed extinct by IUCN when there is no reasonable doubt that the last individual has died.

Extinct in the wild a species is deemed Extinct in the Wild by IUCN when it is known only to survive in cultivation, in captivity or as a naturalized population (or populations) well outside the past range.

Food web the network of connections between species that represents their feeding relationships (who eats, or is eaten by, whom).

Genetic diversity a measure of the genetic differences among individuals of the same species based upon how many alleles are present and their relative frequencies.

Genetic rescue a conservation strategy that reduces the dangers of inbreeding that arise in small populations by introducing new, unrelated individuals from elsewhere.

Greenhouse gases Certain gases in the atmosphere contribute to its insulating properties and therefore contribute to the greenhouse effect. These include carbon dioxide (CO_2), methane (CH_3) and water vapour.

Heat island effect the tendency for city centres to be warmer than the surrounding countryside as a result of the heat-storing capacity of the materials used in buildings and the waste heat from activities of humans.

Herbivore an animal that eats only plants.

IUCN the International Union for the Conservation of Nature is a United Nations agency responsible for evaluating extinction threats to species that are placed upon its Red List.

Least concern a species is deemed to be of Least Concern by IUCN when it has been evaluated against the criteria and does not qualify for Critically Endangered, Endangered, Vulnerable or Near Threatened status. Widespread and abundant taxa are included in this category. These species are not on Red Lists.

Living Planet Index (LPI) an index based upon estimates of the abundance of 5,000 populations of wild animals belonging to over 1,500 species around the world that is designed to track the changing condition of nature.

Megafauna giant animals weighing more than 44 kg (100 lb), many of which went extinct when they came into contact with humans.

Microbe a single-celled organism. Most belong to the archaea or the bacteria.

Mitochondria (singular – mitochondrion). These small bodies found in the cells of all eukaryotes are the power houses that drive metabolism. Mitochondria contain DNA, which is a sign of their origins as once free-living bacteria. The sequence of mitochondrial DNA can be used to trace the evolutionary relationships among the eukaryote species in whose cells they live and are transmitted.

Natural selection the mechanism discovered by Charles Darwin and Alfred Russel Wallace by which adaptation evolves. Natural selection is the process by which inherited traits that increase the number of descendants an individual will leave become increasingly frequent in succeeding generations.

Near Threatened a species is deemed Near Threatened by IUCN when it has been evaluated against the criteria but does not qualify for Critically Endangered, Endangered or Vulnerable now, but is close to qualifying for or is likely to qualify for a threatened category in the near future.

Ocean acidification The increase in the concentration of carbon dioxide (CO_2) in the atmosphere results in an increase in the amount of this gas that dissolves in seawater. When dissolved in seawater, CO_2 reacts with water molecules and liberates hydrogen ions (H^+) that increase the acidity of the oceans.

Reconciliation ecology a scientific approach to accommodating wild species within human-modified or human-occupied landscapes.

Red List the global list of threatened species compiled by IUCN. Sub-lists for particular groups of organisms or individual countries are called Red Data Books.

Redundancy a measure of the replaceability of species in ecosystems.

Resilience the ability of an ecosystem (or population) to bounce back after exploitation.

Resistance a resistant ecosystem is one that does not change under the impact of exploitation.

Speciation the process by which new species evolve.

Species according to the biological definition, organisms belong to the same species if two members can interbreed in the wild and produce viable offspring that can themselves breed.

Sustainable in the context of biodiversity this means that exploitation of a species should be limited to a level that does not endanger the long-term viability of its population.

Symbiosis ecological relationships that involve different species living very closely together, often one being inside the other. Lichens, for example are the result of symbiosis between a fungus and an alga. Mitochondria and chloroplasts originated as symbionts of eukaryotic cells, but the relationship is now so close that they are not regarded as separate species from their eukaryote hosts.

Threatened a species is deemed by IUCN to be Threatened when the risk of extinction is sufficient to place it in any of the categories they define as Vulnerable, Endangered or Critically Endangered.

Top predator a large predator such as shark, wolf or lion at the very top of its food web.

Tragedy of the commons describes the situation in which resources that have no owner, or where there is no limit to an individual's right to exploit common resources, can lead to environmental degradation through practices such as over-grazing of livestock on common land and over-fishing of the seas.

Transect a line used to sample along an environmental gradient, such as from urban to rural areas.

Tree of life the representation of the evolutionary relationships among all living (and extinct) species and groups drawn in the form of a branching tree.

UNESCO United Nations Educational Scientific and Cultural Organization.

Vulnerable a species is deemed Vulnerable by IUCN when the best available evidence indicates that it is facing a high risk of extinction in the wild.

Web of nature the network of ecological connections between species that is forged by their feeding relationships and other kinds of inter-dependency.

Sources

CHAPTER 1 *Biodiversity and us*

Diamond, J. (2005) *Collapse. How societies choose to fail or survive.* Allen Lane, London.

Flenley, J. and Bahn, P. (2002) *The Enigmas of Easter Island.* Oxford University Press, Oxford.

Hails, C., Humphrey, S., Loh, J. and Goldfinger, S. (2008) *Living Planet Report 2008.* WWF International, Gland.

Kareiva, P., Watts, S., McDonald, R. and Boucher, T. (2007) Domesticated nature: shaping landscapes and ecosystems for human welfare. *Science,* **316**, 1866-1869.

Salmond, A. (2004) *The Trial of the Cannibal Dog. Captain Cook in the South Seas.* Penguin, London.

CHAPTER 2 *Life's long chequered career*

Alroy, J. *et al.* (2008) Phanerozoic trends in the global sdiversity of marine invertebrates. *Science,* **321**, 97–100.

Brocks, J.J. *et al.* (2003) A reconstruction of Archean biological diversity based on molecular fossils from the 2.78 to 2.45 billion-year-old Mount Bruce Supergroup, Hamersley Basin, Western Australia. *Geochimica et Cosmochimica Acta,* **67**, 4321–4335.

Budd, G.E. (2008) The earliest fossil record of the animals and its significance. *Philosophical Transactions of the Royal Society, series B,* **363**, 1425–1434.

Butterfield, N.J. (2000) *Bangiomorpha pubescens* n. gen., n. sp.: implications for the evolution of sex, multicellularity, and the Mesoproterozoic/Neoproterozoic radiation of eukaryotes. *Paleobiology,* **26**, 386–404.

Butterfield, N.J. (2001) Cambrian food webs. In: Briggs D.E.G, Crowther P.R, editors. *Palaeobiology II.* Blackwell Science, Oxford, UK. pp. 40–43.

Futuyma, D.J. (2009) *Evolution,* second edition. Sinauer Associates, Sunderland MA.

Hagadorn, J.W. *et al.* (2006) Cellular and subcellular structure of Neoproterozoic embryos. *Science,* **314**, 291–294.

Harper, E.M. (2003) The Mesozoic marine revolution. In: Kelley P.H., Kowalewski, M. and Hanson, T.A., editors. *Predator–Prey Interactions in the Fossil Record.* Kluwer/Plenum, New York. pp. 433–455.

Knoll, A.H. (2003) *Life on a Young Planet: The First Three Billion Years of Evolution on Earth.* Princeton University Press, Princeton, NJ.

Rivera, M.C. and Lake, J.A. (2004) The ring of life provides evidence for a genome fusion origin of eukaryotes. *Nature,* **431**, 152–155.

Sheldon, P.R. *Life in the Palaeozoic.* Open-source introduction to Palaeozoic life, including the invasion of the land: http://openlearn.open.ac.uk/mod/resource/view.php?id=171763 (accessed 13.08.09)

CHAPTER 3 *Evolution's atlas*

Berry, A., ed. (2002) *Infinite Tropics. An Alfred Russel Wallace Anthology.* Verso, London and New York.

Bush, M.B. and Flenley, J.R., eds. (2007) *Tropical Rainforest Responses to Climatic Change.* Springer/Praxis, Chichester.

Da Silva, J.M.C., Rylands, A.B. and Da Fonseca, G.A.B. 2005. The fate of the Amazonian areas of endemism. *Conservation Biology,* **19**, 689–694.

Haffer, J. (1969) Speciation in Amazonian forest birds. *Science,* **165**, 131–137.

Hooghiemstra, H. *et al.* (2006) Late Quaternary palynology in marine sediments: a synthesis of the understanding of pollen distribution patterns in the NW African setting. *Quaternary International,* **148**, 29–44.

Kreft, H. and Jetz, W. (2007) Global patterns and determinants of vascular plant diversity. *Proceedings of the National Academy of Sciencesof the United States of America,* **104**, 5925–5930.

Macey, J.R. *et al.* (1998) Phylogenetic relationships among agamid lizards of the *Laudakia caucasia* species group: testing hypotheses of biogeographic fragmentation and an area cladogram for the Iranian plateau. *Molecular Phylogenetics and Evolution,* **10**, 118–131.

Mann, C.C. (2006) *1491: New revelations of the Americas before Columbus.* Vintage Books, New York.
Myers, N. *et al.* (2000) Biodiversity hotspots for conservation priorities. *Nature,* **403**, 853–858.

Myster, R.W. (2009) Plant communities of Western Amazonia *Botanical Review,* **75**, 1–21.

Parmentier, I. *et al.* (2007) The odd man out? Might climate explain the lower tree-diversity of African rain forests relative to Amazonian rain forests? *Journal of Ecology,* **95**, 1058–1071.

Rasanen, M.E. Linna, A.M., Santos, J.C.R. and Negri, F.R. 1995 Late Miocene tidal deposits in the Amazon foreland basin. *Science* 269, 386-390.

Roosevelt, A.C. *et al.* (1996) Paleoindian cave dwellers in the Amazon: the peopling of the Americas. *Science,* **272**, 373–384.

Verboom, G.A. *et al.* (2009) Origin and diversification of the Greater Cape flora: ancient species repository, hot-bed of recent radiation, or both? *Molecular Phylogenetics and Evolution,* **51**, 44–53.

White, K. and Mattingly, D.J. (2006) Ancient lake of the Sahara. *American Scientist.* **94**, 58–65.

CHAPTER 4 *What is a species?*

Ainouche, M.L. *et al.* (2009) Hybridization, polyploidy and invasion: lessons from *Spartina* (Poaceae). *Biological Invasions,* **11**, 1159–1173.

Brown, D.M. *et al.* (2007) Extensive population genetic structure in the giraffe. *BMC Biology,* **5**, 57.

Buckley, T.R., Attanayake, D. and Bradler, S. (2009) Extreme convergence in stick insect evolution: phylogenetic placement of the Lord Howe Island tree lobster. *Proceedings of the Royal Society, series B,* **276**, 1055–1062.

Darwin, C.R. (1859) *The Origin of Species,* first edition (reprinted). Murray, London.

Haas, F. and Brodin, A. (2005) The crow *Corvus corone* hybrid zone in Southern Denmark and northern Germany. *Ibis,* **147**, 649–656.

Lincoln, R. (1979) *British Marine Amphipods.* BMNH, London.

Nowak, R.M. (1992) The red wolf is not a hybrid. *Conservation Biology,* **6**(4), 593–595.

Owen, D.F. and Smith, D.A.S. (1993) Spot variation in *Maniola jurtina* (L.) (Lepidoptera: Satyridae) in Southern Portugal and a comparison with the Canary Islands. *Biological Journal of the Linnean Society*, **49**, 355–365.

Saino, N. and Villa, S. (1992) Pair composition and reproductive success across a hybrid zone of carrion crows and hooded crows. *The Auk*, **109** (3), 543–555.

Vieites, D.R. *et al.* (2009) Vast underestimation of Madagascar's biodiversity evidenced by an integrative amphibian inventory. *Proceedings of the National Academy of Sciencesof the United States of America*, **106**(20), 8267–8272.

Wallace, A.R. (1880) *Island Life*. Macmillan, London.

Wayne, B. (1995) Red wolves: to conserve or not to conserve. *Canid News*, **3**.

CHAPTER 5 *Genes, genes, genes*

Balanyá, J. *et al.* (2007) Global genetic change tracks global climate warming in *Drosophila subobscura*. *Science*, **313**, 1773–1775.

Beadle, G. (1939) Teosinte and the origin of maize. *Journal of Heredity*, **30**, 245–247.

Dever, J.A. *et al.* (2002) Genetic diversity, population subdivision, and gene flow in Morelet's crocodile (*Crocodylus moreletti*) from Belize, Central America. *Copeia*, **4**, 1078–1091.

Frankham, R., Ballou, J.D. and Briscoe, D.A. (2002) *Introduction to Conservation Genetics*. Cambridge University Press, Cambridge.

Freeland, J.R. (2005) *Molecular Ecology*. Wiley & Sons, Chichester.

Golden Lion Tamarin Conservation Program. http://nationalzoo.si.edu/ConservationAndScience/EndangeredSpecies/GLTProgram/ZooLife/CurrentStatus.cfm (accessed 25.09.09).

Hewitt, G.M. (1999) Post-glacial re-colonization of European biota. *Biological Journal of the Linnean Society*, **68**, 87–112.

Land, E.D. and Lacy, R.C. (2000) Introgression level achieved through Florida panther genetic restoration. *Endangered Species Update*, **17**, 100–105.

Madsen, T. *et al.* (1999) Conservation biology – restoration of an inbred adder population. *Nature*, **402**, 34–35.

Nielsen, R.K., Pertoldi, C. and Loeschcke, V. (2007) Genetic evaluation of the captive breeding program of the Persian wild ass. *Journal of Zoology*, **272**, 349–357.

Reale, D., McAdam, A.G., Boutin, S. and Berteaux, D. (2003) Genetic and plastic responses of a northern mammal to climate change. *Proceedings of the Royal Society London, Series B*, **270**, 591–596.

Roelke, M.E., Martenson, J.S. and O'Brien, S.J. (1993) The consequences of demographic reduction and genetic depletion in the endangered Florida panther. *Current Biology*, **3**, 340–350.

Russello, M.A. and Amato, G. (2004) *Ex situ* population management in the absence of pedigree information. *Molecular Ecology*, **13**, 2829–2840.

Saccheri, I. *et al.* (1998) Inbreeding and extinction in a butterfly metapopulation. *Nature*, **392**, 491–494

Tzika, A. *et al.* (2009) Molecular genetic analysis of a captive-breeding program: the vulnerable endemic Jamaican yellow boa. *Conservation Genetics*, **10**, 69–77.

CHAPTER 6 *The web of nature*

Armbrust, E.V. (2009) The life of diatoms in the world's oceans. *Nature*, **459**, 185–192.

Barrios, E. (2007) Soil biota, ecosystem services and land productivity. *Ecological Economics*, **64**, 269–285.

Bonan, G.B. (2008) Forests and climate change: forcings, feedbacks, and the climate benefits of forests. *Science*, **320**, 1444–1449.

Coleman, D.C. (2008) From peds to paradoxes: linkages between soil biota and their influences on ecological processes. *Soil Biology and Biochemistry*, **40**, 271–289.

DeLong, E.F. (2009) The microbial ocean from genomes to biomes. *Nature*, **459**, 200–206.

Fierer, N., Breitbart, M., Nulton, J., Salamon, P., Lozupone, C., Jones, R., Robeson, M., Edwards, R.A., Felts, B., Rayhawk, S., Knight, R., Rohwer, F. and Jackson, R.B. (2007) Metagenomic and small-subunit rRNA analyses reveal the genetic diversity of bacteria, archaea, fungi, and viruses in soil. *Applied and Environmental Microbiology*, **73**, 7059–7066.

Fuhrman, J.A. (2009) Microbial community structure and its functional implications. *Nature*, **459**, 193–199.

King, R.A., Tibble, A.L. and Symondson, W.O.C. (2008) Opening a can of worms: unprecedented sympatric cryptic diversity within British lumbricid earthworms. *Molecular Ecology*, **17**, 4684–4698.

Nichols, E. *et al.* (2008) Ecological functions and ecosystem services provided by Scarabaeinae dung beetles. *Biological Conservation*, **141**, 1461–1474.

Rohwer, F. and Thurber, R.V. (2009) Viruses manipulate the marine environment. *Nature*, **459**, 207–212.

Strong, A., Chisholm, S., Miller, C. and Cullen, J. (2009) Ocean fertilization: time to move on. *Nature*, **461**, 347–348.

CHAPTER 7 *What has biodiversity ever done for us?*

Allison, S.D. and Martiny, J.B.H. (2008) Resistance, resilience, and redundancy in microbial communities. *Proceedings of the National Academy of Sciences of the United States of America*, **105**, 11512–11519.

Barrios, E. (2007) Soil biota, ecosystem services and land productivity. *Ecological Economics*, **64**, 269–285.

Beschta, R.L. and Ripple, W.J. (2006) River channel dynamics following extirpation of wolves in northwestern Yellowstone National Park, USA. *Earth Surface Processes and Landforms*, **31**, 1525–1539.

Casini, M., *et al.* (2009) Trophic cascades promote threshold-like shifts in pelagic marine ecosystems. *Proceedings of the National Academy of Sciences of the United States of America*, **106**, 197–202.

Dobson, A. (2009) Food-web structure and ecosystem services: insights from the Serengeti. *Philosophical Transactions of the Royal Society B* **364**, 1665–1682.

Freinkel, S. (2007) *American Chestnut*. California University Press, Berkley.

Leopold, A. (1949) A sand county almanac and other writings. Oxford University Press, New York.

Locutus of Borg (1990) *Star Trek: The Next Generation*, episode *The Best of Both Worlds* http://en.wikipedia.org/wiki/Resistance_is_futile#cite_note-Best-3 (accessed 05.06.09).

Mando, A. and Stroosnijder, L. (1999) The biological and physical role of mulch in the rehabilitation of crusted soil in the Sahel. *Soil Use and Management*, **15**, 123–127.

Murphy, S., Tone, J.W. and Schwartzberg, P. (1995) Land Acquisition for Water Quality Protection: New York City and the Catskills Watershed System. Integrated Watershed Management – A New Paradigm for Water Management? (Ward, R.C., editor). Universities Council on Water Resources www.ucowr.siu.edu/updates/pdf/V100_A9.pdf(accessed 05.06.09).

Pain, D.J. *et al.* (2008) The race to prevent the extinction of South Asian vultures. *Bird Conservation International*, **18**, S30–S48.
Resilience Alliance (2009) Thresholds and Alternate States in Ecological and Social–Ecological Systems. A Resilience Alliance / Santa Fe Institute Database. www.resalliance.org(accessed 12.06.09).

Ripple, W.J. and Beschta, R.L. (2007) Restoring Yellowstone's aspen with wolves. *Biological Conservation*, **138**, 514–519.

Sergio, F. *et al.* (2008) Top predators as conservation tools: ecological rationale, assumptions, and efficacy. *Annual Review of Ecology Evolution and Systematics*, **39**, 1–19.

Thompson, R. and Starzomski, B.M. (2007) What does biodiversity actually do? A review for managers and policy makers. *Biodiversity and Conservation*, **16**, 1359–1378.

van der Heijden, M.G.A., Bardgett, R.D. and van Straalen, N.M. (2008) The unseen majority: soil microbes as drivers of plant diversity and productivity in terrestrial ecosystems. *Ecology Letters*, **11**, 296–310.

Worm, B. *et al.* (2006) Impacts of biodiversity loss on ocean ecosystem services. *Science*, **314**, 787–790.

CHAPTER 8 *Valued landscapes*

Hadfield, M. (1985) *A History of British Gardening*. Penguin, London.

Nash, R. ed. (1990) *American Environmentalism: Readings in Conservation History*. McGraw Hill, Columbus, OH.

Oelschlaeger, M. (1991) *The Idea of Wilderness: from Pre-history to the Age of Ecology*. Yale University Press, New Haven CT.

Pepper, D. (1996) *Modern Environmentalism: an Introduction*. Routledge, New York.

Van Koppen, C. and Markham, W. (2007) *Protecting Nature: Organizations and Networks in Europe and the USA*. Edward Elgar, Cheltenham.

World Commission on Environment and Development (1987) *Our Common Future (The Brundtland Report)*. Oxford University Press, Oxford.

Wulf, A. (2008) *The Brother Gardeners: Botany, Empire and the Birth of an Obsession*. Heinemann, Portsmouth NH.

CHAPTER 9 *Nature in the city*

Ausden, M. (2007) *Habitat Management for Conservation*. Oxford University Press, Oxford.
Barker G. and Graf, A. (1989) *Urban Wildlife Now – No. 3 Principles for Nature Conservation in Towns and Cities* JNCC Peterborough.

Benton-Short, L. and Short, J.R .(2008) Cities and Nature. Routledge Taylor & Francis Group, London and New York.

Carr, S. and Lane, A. (1993) *Practical Conservation: Urban Habitats*. Hodder and Stoughton, London.

Gaston, K.J., *et al.* (2004) Garden Wildlife – The BUGS project. *British Wildlife*, **16**, 1–9.

Gibb, H. and Hochuli, D.F (2002) Habitat fragmentation in an urban environment: large and small fragments support different arthropod assemblages. *Biological Conservation*, **106**, 91–100.

Haskins, L. (2000) Heathlands in an urban setting – effects of urban development on heathlands of south-east Dorset. *British Wildlife*, **11**, 229–237.

Marzluff, J.M. *et al.* (2008) *Urban Ecology. An International Perspective on the Interaction between Human and Nature*. Springer, New York.

Richardson, D.F.H. (1992) *Pollution Monitoring with Lichens. Naturalists' Handbooks no. 19*. Richmond Publishing Co Ltd., Slough.

Toms, M. (2007) Are gardens good for birds or birdwatchers? *British Wildlife*, **19**, 77–83

CHAPTER 10 *Life in freshwater*

AAAS (2000) *Atlas of Population and Environment*. University of California Press, Berkeley CA.

Allan, J.D. *et al.* (2005) Overfishing of inland waters. *BioScience*, **55**, 1041–1051.

Buch, A. and Dixon, A.B. (2009) South Africa's Working for Water programme: searching for win-win outcomes for people and the environment. *Sustainable Development*, **17**, 129–149.

Collins, J.P. and Halliday, T. (2005) Forecasting changes in amphibian biodiversity: aiming at a moving target. Philosophical Transactions of the Royal Society B, **360**, 309–314.

Conservation International (2009) *Annual Report for 2008*. Conservation International, Arlington, VA.

Dudgeon, D. *et al.* (2006) Freshwater biodiversity: importance, threats, status and conservation challenges. *Biological Review* **81**, 163–182.

Halliday T. and Davey, B. (2007) *Water and Health in an Overcrowded World*. SDK125, Book 1. Oxford University Press, Oxford. p. 56.
Hayes, T.B. et al. (2002a) Hermaphroditic, demasculinized frogs after exposure to the herbicide atrazine at low ecologically relevant doses. *Proceedings of the National Academy of Sciences of the United States of America*, **99**, 5476–5480.

Hayes, T.B. *et al.* (2002b) Feminization of male frogs in the wild. *Nature*, **419**, 895–896.

Houghton, J. (2004) *Global Warming. The Complete Briefing*, third edition. Cambridge University Press, Cambridge.

Lean, G. and Hinrichsen, D. (1992) *Atlas of the Environment*, second edition. World Wildlife Fund, Harper Perennial, New York.

Loh, J. *et al.* (2005) *Living Planet Report 2004*. WWF, Gland.

Marais, C. and Wannenburgh, A.M. (2008) Restoration of water resource (natural capital) through the clearing of invasive alien plants from riparian areas in South Africa – costs and water benefits. *South African Journal of Botany*, **74**, 526–537.

Revenga, C., Campbell, I., Abell, R., de Villiers, P. and Bryer, M. (2005) Prospects for monitoring freshwater ecosystems towards the 2010 targets. *Philosophical Transactions of the Royal Society, series B*, **360**, 397–413.

Revenga, C. and Mock, G. (2001) *Freshwater Biodiversity in Crisis*. Earth Trends (World Resources Institute). www.earthtrends.wri.org

UNESCO (2003) *Water for People, Water for Life*. The United Nations World Water Development Report. www.unesco.org/water/wwap

van Jaarsveld, A.S. *et al.* (2005) Measuring conditions and trends in ecosystem services at multiple scales: the Southern African Millennium Ecosystem Assessment (SAfMA) experience. Philosophical Transactions of the Royal Society, series B, **360**, 425–441.

Williams, P. *et al.* (2004) Comparative biodiversity of rivers, streams, ditches and ponds in agricultural landscape in southern Britain. *Biological Conservation*, **115**, 329–341.

CHAPTER 11 *Going, going gone; the sixth extinction*

AmphibiaWeb. www.amphibiaweb.org

Cleaveland, S. *et al.* (2007) The conservation relevance of epidemiological research into carnivore viral diseases in the Serengeti. *Conservation Biology*, **21**, 612–622.

Conservation International. hotspot web pages. http://www.conservation.org/explore/priority_areas/hotspots/(accessed 13.09.09).
Daszak, P., Cunningham, A.A. and Hyatt, A.D. (2000) Emerging infectious diseases of wildlife – threats to biodiversity and human health. *Science*, **287**, 443–449.

Drost, C.A. and Fellers, G.M. (1996) Collapse of a regional frog fauna in the Yosemite area of the California Sierra Nevada, USA. *Conservation Biology*, **10**, 414–425.

Eldridge, N. (2001) The sixth extinction. www.actionbioscience.org/newfrontiers/eldredge2.html

Halliday, T. (1978) *Vanishing Birds*. Sidgwick & Jackson, London.

Halliday, T. (2007) Amphibian decline. In: *Encyclopedia of Life Sciences*. Wiley, Chichester.

Hansen, D.M. and Galetti, M. (2009) The forgotten megafauna. *Science*, **324**, 42.

Hayes, T.B. *et al.* (2002) Feminization of male frogs in the wild. *Nature*, **419**, 895–896.

Houlahan, J.E. *et al.* (2000) Quantitative evidence for global amphibian population declines. *Nature*, **404**, 752–754.

IUCN Red List Categories (2001) http://www.iucnredlist.org/static/categories_criteria_3_1[Accessed 06.09.09]

Jablonski, D. (1995) Extinctions in the fossil record. In: Lawton, J.H. and May, R.M. editors, *Extinction Rates*. Oxford University Press, Oxford. pp.25–44.

Kiesecker, J.M. (2002) Synergism between trematode infection and pesticide exposure: a link to amphibian limb deformities in nature? *Proceedings of the National Academy of Sciences of the United State of America*, **99**, 9900–9904.

MacLeod, N. (2005) Extinction. In: *Encyclopedia of Life Sciences*. Wiley, Chichester.

McAloose, D. and Newton, A.L. (2009) Wildlife cancer: a conservation perspective. *Nature Reviews Cancer*, **9**, 517–526.

McKenzie, V.J. and Townsend, A.R. (2007) Parasitic and infectious disease responses to changing global nutrient cycles. *EcoHealth*, **4**, 384–396.

McKinney, M.L. (1999) High rates of extinction and threat in poorly studied taxa. *Conservation Biology*, **13**, 1273–1281.

McKinney, M.L. and Lockwood, J.L. (1999) Biotic homogenization: a few winners replacing many losers in the next mass extinction. *Trends in Ecology and Evoloution*, **14**(11), 450–453.
Stein, B.A., Kutner, L.S. and Adams, J.S. editors. (2000) *Precious Heritage: The Status of Biodiversity in the United States*. Oxford University Press, New York.

Stuart, S., Chanson, J.S., Cox, N.A. *et al.* (2004) Status and trends of amphibian declines and extinctions worldwide. *Science*, **306**, 1783–1786.

Wake, D.B. and Vredenburg, V.T. (2008) Are we in the midst of the sixth mass extinction? The view from the world of amphibians. *Proceedings of the National Academy of Sciences of the United States of America*., **105** (Suppl. 1), 11466–11473.

World Resources Institute (2008) *A Guide to World Resources 2008: Roots of Resilience Growing the Wealth of the Poor*. World Resources Institute, Washington, DC.

WWF (2004) *The Living Planet Report*. World Wildlife Fund, Gland.

CHAPTER 12 *The exhaustible sea*

Barrett, J.H., Locker, A.M. and Roberts, C.M. (2004) The origins of intensive marine fishing in Medieval Europe: the English evidence. *Proceedings of the Royal Society series B* **271**, 2417–2421.

Beddington, J.R., Agnew, D.J. and Clark, C.W. (2007) Current problems in the management of marine fisheries. *Science*, **316**, 1713–1716.

Mora, C. *et al.* (2009) Management effectiveness of the world's marine fisheries. *PLoS Biology*, **7**, 1–11.

Pauly, D., Alder, J., Bennett, E., Christensen, V., Tyedmers P. and Watson R. (2003) The future for fisheries. *Science*, **302**, 1359–1361.

Pinnegar, J.K. and Engelhard, G.H. (2008) The shifting baseline phenomenon: a global perspective. *Reviews in Fish Biology and Fisheries*, **18**, 1–16.

Pisco (Partnership for Interdisciplinary Studies of Coastal Oceans) *The Science of Marine Reserves*. www.piscoweb.org

Roberts, C.M. *et al.* (2001) Effects of marine reserves on adjacent fisheries. *Science*, **294**, 1920–1923.

Roberts, C.M. (2007) *The Unnatural History of the Sea*. Island Press, Washington DC and Gaia Thinking, London

Saenz-Arroyo, A., Roberts, C.M., Torre, J., Carino-Olvera, M. and Enriquez Andrade, R.R. (2005) Rapidly shifting environmental baselines among fishers of the Gulf of California. *Proceedings of the Royal Society series B*, **272**, 1957–1962.

Worm, B. *et al.* (2006) Impacts of biodiversity loss on ocean ecosystem services. *Science*, **314**, 787–790.

Worm, B. *et al.* (2009) Rebuilding global fisheries. *Science*, **325**, 578–585.

CHAPTER 13 *Climate change*

Adam, D. (2009) How global warming sealed the fate of the world's coral reefs. *The Guardian*. London.

Baskett, M.L., Gaines, S.D. and Nisbet, R.M. (2009) Symbiont diversity may help coral reefs survive moderate climate change. *Ecological Applications*, **19**, 3–17.

Beckage, B., *et al.* (2008) A rapid upward shift of a forest ecotone during 40 years of warming in the Green Mountains of Vermont. *Proceedings of the National Academy of Sciences of the United States of America*, **105**, 4197–4202.

Both, C., Bouwhuis, S., Lessells, C.M. and Visser, M.E. (2006) Climate change and population declines in a long-distance migratory bird. *Nature*, **441**, 81–83.

Both, C., van Asch, M., Bijlsma, R.G., van den Burg, A.B. and Visser, M.E. (2009) Climate change and unequal phenological changes across four trophic levels: constraints or adaptations? *Journal of Animal Ecology*, **78**, 73–83.

Bradley, B.A., Oppenheimer, M. and Wilcove, D.S. (2009) Climate change and plant invasions: restoration opportunities ahead? *Global Change Biology*, **15**, 1511–1521.

Carpenter, K.E. *et al.* (2008) One third of reef-building corals face elevated extinction risk from climate change and local impacts. *Science*, **321**, 560–563.

Chave, J. *et al.* (2008) Assessing evidence for a pervasive alteration in tropical tree communities. *PLoS Biology*, **6**, 455–462.

Chen, I.C. *et al.* (2009) Elevation increases in moth assemblages over 42 years on a tropical mountain. *Proceedings of the National Academy of Sciences of the United States of America*, **106**, 1479–1483.

Dunbar, H.E., Wilson, A.C.C., Ferguson, N.R. and Moran, N.A. (2007) Aphid thermal tolerance is governed by a point mutation in bacterial symbionts. *PLoS Biology*, **5**, 1006–1015.

Fitter, A.H. and Fitter, R.S.R. (2002) Rapid changes in flowering time in British plants. *Science*, **296**, 1689–1691.

Gange, A.C., Gange, E.G., Sparks, T.H. and Boddy, L. (2007) Rapid and recent changes in fungal fruiting patterns. *Science*, **316**, 71.

Gardner, T.A., Barlow, J., Chazdon, R., Ewers, R.M., Harvey, C.A., Peres, C.A. and Sodhi, N.S. (2009) Prospects for tropical forest biodiversity in a human-modified world. *Ecology Letters*, **12**, 561–582.

Hamann, A. and Wang, T. (2006) Potential effects of climate change on ecosystems and tree species distribution in British Columbia. *Ecology*, **87**, 2773–2786.

Hannah, L., Midgley, G., Hughes, G. and Bomhard, B. (2005) The view from the cape. Extinction risk, protected areas, and climate change. *Bioscience*, **55**, 231–242.

Hickling, R., Roy, D.B., Hill, J.K., Fox, R. and Thomas, C.D. (2006) The distributions of a wide range of taxonomic groups are expanding polewards. *Global Change Biology*, **12**, 450–455.

Hoegh-Guldberg, O., Hughes, L., McIntyre, S., Lindenmayer, D.B., Parmesan, C., Possingham, H.P. and Thomas, C.D. (2008) Assisted colonization and rapid climate change. *Science*, **321**, 345–346.

Hole, D.G. *et al.* (2009) Projected impacts of climate change on a continent-wide protected area network. *Ecology Letters*, **12**, 420–431.

Jones, A.M., Berkelmans, R., van Oppen, M.J.H., Mieog, J.C. and Sinclair, W. (2008) A community change in the algal endosymbionts of a scleractinian coral following a natural bleaching event: field evidence of acclimatization. *Proceedings of the Royal Society B-Biological Sciences*, **275**, 1359–1365.

Karl, T.R., Melillo, J.M. and Peterson, T.C. (2009) *Global Climate Change Impacts in the United States*. Cambridge University Press, Cambridge.

Lawler, J.J. *et al.* (2009) Projected climate-induced faunal change in the Western Hemisphere. *Ecology*, **90**, 588–597.

Lenoir, J., Gegout, J.C., Marquet, P.A., de Ruffray, P. and Brisse, H. (2008) A significant upward shift in plant species optimum elevation during the 20th century. *Science*, **320**, 1768–1771.

Lewis, S.L., Malhi, Y. and Phillips, O.L. (2004) Fingerprinting the impacts of global change on tropical forests. *Philosophical Transactions of the Royal Society of London Series B-Biological Sciences*, **359**, 437–462.

Lewis, S.L. *et al.* (2004) Concerted changes in tropical forest structure and dynamics: evidence from 50 South American long-term plots. *Philosophical Transactions of the Royal Society of London Series B-Biological Sciences*, **359**, 421–436.

Marris, E. (2009) Planting the forest of the future. *Nature*, **459**, 906–908.

Miller-Rushing, A.J. and Primack, R.B. (2008) Global warming and flowering times in Thoreau's concord: a community perspective. *Ecology*, **89**, 332–341.

Moritz, C., Patton, J.L., Conroy, C.J., Parra, J.L., White, G.C. and Beissinger, S.R. (2008) Impact of a century of climate change on small-mammal communities in Yosemite National Park, USA. *Science*, **322**, 261–264.

Nielsen, J.T. and Moller, A.P. (2006) Effects of food abundance, density and climate change on reproduction in the sparrowhawk *Accipiter nisus*. *Oecologia*, **149**, 505–518.

Parmesan, C. and Yohe, G. (2003) A globally coherent fingerprint of climate change impacts across natural systems. *Nature*, **421**, 37–42.

Phillips, O.L. and Gentry, A.H. (1994) Increasing turnover through time in tropical forests. *Science*, **263**, 954–958.

Phillips, O.L. *et al.* (2004) Pattern and process in Amazon tree turnover, 1976–2001. *Philosophical Transactions of the Royal Society of London Series B-Biological Sciences*, **359**, 381–407.

Purse, B.V. *et al.* (2007) Incriminating bluetongue virus vectors with climate envelope models. *Journal of Applied Ecology*, **44**, 1231–1242.

Sparks, T.H. and Carey, P.D. (1995) The responses of species to climate over two centuries: an analysis of the Marsham phenological record, 1736–1947. *Journal of Ecology*, **83**, 321–329.

Thomas, C.D. *et al.* (2004) Extinction risk from climate change. *Nature*, **427**, 145–148.

Thoreau, H.D. (1854) *Walden; or, Life in the Woods*. Castle Books, Edison, NJ.

Thuiller, W., Broennimann, O., Hughes, G., Alkemade, J.R.M., Midgley, G.F. and Corsi, F. (2006) Vulnerability of African mammals to anthropogenic climate change under conservative land transformation assumptions. *Global Change Biology*, **12**, 424–440.

UNEP-WCMC (2006) *In the Front Line: Shoreline Protection and Other Ecosystem Services from Mangroves and Coral Reefs*, 33 pp. UNEP-WCMC, Cambridge.

van Mantgem, P.J. *et al.* (2009) Widespread increase of tree mortality rates in the Western United States. *Science*, **323**, 521–524.

Wilson, R.J. *et al.* (2005) Changes to the elevational limits and extent of species ranges associated with climate change. *Ecology Letters*, **8**, 1138–1146.

CHAPTER 14 *Invasion plant Earth*

Baskin, Y. (2002) *A Plague of Rats and Rubber-vines. The Growing Threat of Species Invasions*. Island Press, Washington D.C.

Clavero, M. and Garcia-Berthou, E. (2005) Invasive species are a leading cause of animal extinctions. *Trends in Ecology and Evolution*, **20**, 110–110.

Galil, B.S. (2007) Loss or gain? Invasive aliens and biodiversity in the Mediterranean Sea. *Marine Pollution Bulletin*, **55**, 314–322.

Garber, S.D. (1998) *The Urban Naturalist*. Dover Publications, Mineola NY.

Gosling, L.M. and Baker, S.J. (1989) The eradication of muskrats and coypus from Britain. *Biological Journal of the Linnean Society*, **38**, 39–51.

Gurnell, J. *et al.* (2006) Squirrel poxvirus: landscape scale strategies for managing disease threat. *Biological Conservation*, **131**, 287–295.

Jager, H., Tye, A. and Kowarik, I. (2007) Tree invasion in naturally treeless environments: impacts of quinine (*Cinchona pubescens*) trees on native vegetation in Galapagos. *Biological Conservation*, **140**, 297–307.

Koenig, W.D. (2003) European starlings and their effect on native cavity-nesting birds. *Conservation Biology*, **17**, 1134–1140.

Letnic, M., Webb, J.K. and Shine, R. (2008) Invasive cane toads (*Bufo marinus*) cause mass mortality of freshwater crocodiles (*Crocodylus johnstoni*) in tropical Australia. *Biological Conservation*, **141**, 1773–1782.

Low, T. (1999) *Feral Future*. Vintage, Melbourne.

Mitchell, C.E. and Power, A.G. (2003) Release of invasive plants from fungal and viral pathogens. *Nature*, **421**, 625–627.

Oliveras, J., Bas, J.M. and Gomez, C. (2007) A shift in seed harvesting by ants following Argentine ant invasion. *Vie Et Milieu-Life and Environment*, **57**, 79–85.

Possingham, H., Jarman, P. and Kearns, A. (2003) *Independent Review of Western Shield. Report of the Review Panel*. Department of Conservation and Land Management, Perth.

Ricciardi, A., Neves, R.J. and Rasmussen, J.B. (1998) Impending extinctions of North American freshwater mussels (Unionoida) following the zebra mussel (*Dreissena polymorpha*) invasion. *Journal of Animal Ecology*, **67**, 613–619.

Rowles, A.D. and O'Dowd, D.J. (2009) Impacts of the invasive Argentine ant on native ants and other invertebrates in coastal scrub in south-eastern Australia. *Australian Ecology*, **34**, 239–248.

Schloesser, D.W. *et al.* (2006) Extirpation of freshwater mussels (Bivalvia : Unionidae) following the invasion of dreissenid mussels in an interconnecting river of the Laurentian Great Lakes. *American Midland Naturalist*, **155**, 307–320.

Shakespeare, W. (2006) *Complete Works of William Shakespeare*. RSC edition. Macmillan, Basingstoke.

Smith, K.W. (2005) Has the reduction in nest-site competition from Starlings *Sturnus vulgaris* been a factor in the recent increase of Great Spotted Woodpecker *Dendrocopos major* numbers in Britain? *Bird Study*, **52**, 307–313.

Strayer, D.L. (2009) Twenty years of zebra mussels: lessons from the mollusk that made headlines. *Frontiers in Ecology and the Environment*, **7**, 135–141.

Strayer, D.L. and Malcom, H.M. (2007) Effects of zebra mussels (*Dreissena polymorpha*) on native bivalves: the beginning of the end or the end of the beginning? *Journal of the North American Benthological Society*, **26**, 111–122.

Vogel, V., Pedersen, J.S., d'Ettorre, P., Lehmann, L. and Keller, L. (2009) Dynamics and genetic structure of Argentine ant supercolonies in their native range. *Evolution*, **63**, 1627–1639.

CHAPTER 15 *What next for nature?*

Abensperg-Traun, M. (2009) CITES, sustainable use of wild species and incentive-driven conservation in developing countries, with an emphasis on southern Africa. *Biological Conservation*, **142**, 948–963.

Balmford, A., Green, R.E. and Scharlemann, J.P.W. (2005) Sparing land for nature: exploring the potential impact of changes in agricultural yield on the area needed for crop production. *Global Change Biology*, **11**, 1594–1605.

Blanc, J.J. *et al.* (2007) *African Elephant Status Report 2007: An Update from the African Elephant Database*. Occasional paper of the IUCN Species Survival Commission, 284 pp. IUCN, Gland, Switzerland.

Burney, D.A. and Flannery, T.F. (2005) Fifty millennia of catastrophic extinctions after human contact. *Trends in Ecology and Evolution*, **20**, 395–401.

Costanza, R. *et al.* (1997) The value of the world's ecosystem services and natural capital. *Nature*, **387**, 253–260.

Diamond, J. (2008) Palaeontology – the last giant kangaroo. *Nature*, **454**, 835–836.

Ellis, E. (2008) *Anthropocene*. In Cleveland, C.J. editor. *Encyclopedia of Earth*. Environmental Information Coalition, National Council for Science and the Environment, Washington, D.C. http://www.eoearth.org/article/Anthropocene. (accessed 18.09.09).

Ewers, R.M., Scharlemann, J.P.W., Balmford, A. and Green, R.E. (2009) Do increases in agricultural yield spare land for nature? *Global Change Biology*, **15**, 1716–1726.

Goldman, R.L., Tallis, H., Kareiva, P. and Daily, G.C. (2008) Field evidence that ecosystem service projects support biodiversity and diversify options. *Proceedings of the National Academy of Sciences of the United States of America*, **105**, 9445–9448.

Guldemond, R. and Van Aarde, R. (2008) A meta-analysis of the impact of African elephants on savanna vegetation. *Journal of Wildlife Management*, **72**, 892–899.

King, L.E., Lawrence, A., Douglas-Hamilton, I. and Vollrath, F. (2009) Beehive fence deters crop-raiding elephants. *African Journal of Ecology*, **47**, 131–137.

Lemieux, A.M. and Clarke, R.V. (2009) The international ban on ivory sales and its effects on elephant poaching in Africa. *British Journal of Criminology*, **49**, 451–471.

Roca, A.L., Georgiadis, N., Pecon-Slattery, J. and O'Brien, S.J. (2001) Genetic evidence for two species of elephant in Africa. *Science*, **293**, 1473–1477

Stiles, D. (2004) The ivory trade and elephant conservation. *Environmental Conservation*, **31**, 309–321.

Stiles, D. (2009) CITES-approved ivory sales and elephant poaching. *Pachyderm*, **45**, 150 153.

Wasser, S.K. *et al.* (2007) Using DNA to track the origin of the largest ivory seizure since the 1989 trade ban. *Proceedings of the National Academy of Sciences of the United States of America*, **104**, 4228–4233.

World Resources Institute (2008) *A Guide to World Resources 2008: Roots of Resilience Growing the Wealth of the Poor*. World Resources Institute, Washington, DC.

UVW
ST
QR
MN
KL
J
EF
CD